ENVIRONMENTAL PHYSICS SERIES

General Editor: C. R. Bassett, B.Sc.

Other books in the Series are:
*Heating*, by C. R. Bassett and M. D. W. Pritchard
*Lighting*, by D. C. Pritchard

# ENVIRONMENTAL PHYSICS

One of the major problems to be considered by construction engineers is that of environmental design. To this end it is essential to have a sound knowledge of the fundamental principles involved and these may be classified under three main headings:

Heating
Lighting
Acoustics

The three volumes in this series have been written by experts in their own field and incorporate the most modern techniques available and the latest research information. In accordance with national policy, S.I. metric units have been used throughout leading to clarity of expression without the ambiguity associated with traditional units.

The books are suitable for students undertaking all aspects of advanced construction technology on Degree, Higher National Diploma, or Architectural courses.

Environmental Physics:

# Acoustics

B. J. Smith, B.Sc.

Ewell Technical College

American Elsevier Publishing Company, Inc.
New York
1970

AMERICAN ELSEVIER PUBLISHING COMPANY, INC.
52 Vanderbilt Avenue
New York, N.Y. 10017

ISBN 0 582 42003 2

LIBRARY OF CONGRESS CATALOG CARD NUMBER: 73-75108

Set in 10 on 12 pt Times New Roman
Printed in England

# Contents

# Acknowledgements

I gratefully acknowledge permission to include Figs. 4.8 and 5 3, absorption coefficients and T.N.1 by the Building Research Station; Figs. 2.13, 2.14 and 2.15 by H.M.S.O. and Figs. 2.7, 2.12, 4.6, 4.9, 5.4, 5.14, 6.5 and 6.6 together with the section on Industrial Noise Affecting Residential Areas by the British Standards Institution. I should also like to thank Dr. R. W. B. Stephens for permission to include the formula for optimum reverberation time.

I should like to express my gratitude to Mr R. C. Slater for the tremendous help he has given me over many years, and through whose encouragement this book has been possible. I am indebted to Mr C. Bassett for his constructive suggestions in the preparation of the manuscript.

To my wife I am very grateful for her untiring work at the typewriter. Also to my baby daughter Kathryn, who has provided the inspiration for a book on noise control.

Chapter 1

# The Measurement of Sound

Sound is an aural sensation caused by pressure variations in the air which are always produced by some source of vibration. They may be from a solid object or from turbulence in a liquid or gas. These pressure fluctuations may take place very slowly, such as those caused by atmospheric changes, or very rapidly and be in the ultrasonic frequency range. The velocity of sound is independent of the rate at which these pressure changes take place and depends solely on the properties of the air in which the sound wave is travelling.

### Frequency

This is the number of vibrations or pressure fluctuations per second. The unit is the hertz (Hz).

### Wavelength

This is the distance travelled by the sound during the period of one complete vibration.

### Velocity of Sound in Air

Velocity = frequency × wavelength

$$V = f\lambda$$

where $V$ = velocity of sound in m/s
$f$ = frequency in Hz
$\lambda$ = wavelength in metres (m)

It is sufficiently accurate for the purpose of Building Acoustics to consider the velocity of sound to be a constant at 330 m/s. The wavelength of a sound of 20 Hz frequency

$$= \frac{330}{20}\text{ m}$$

$$= 16{\cdot}5\text{ m}$$

The wavelength of a sound of 20 kHz frequency

$$= \frac{330}{20\,000}$$

$$= 0{\cdot}0165\text{ m}$$
$$= 16{\cdot}5\text{ mm}$$

These are the extremes of wavelength for audible sounds.

## Propagation of Sound Waves

Air cannot sustain a shear force so that the only type of wave possible is longitudinal, where the vibrations are in the direction of the motion. This is illustrated in Fig. 1.1. These pressure fluctuations are of a vibrational nature causing the neighbouring air pressure to change but no movement of the air takes place. Air pressure, which can be assumed to be steady, has these pressure fluctuations superimposed upon it.

Reflection of sound takes place when there is a change of medium. The larger the change the greater the amount of reflection and the smaller the transmission. The laws of reflection for sound are similar to those for light:

1. The angle of incidence is equal to the angle of reflection.
2. The incident wave, the reflected wave and the normal all lie in the same plane.

There is a limitation on the first of these. The reflecting surface must have dimensions of at least the same order of size as the wavelength of the sound. If the reflecting object is much smaller than the wavelength then diffraction will take place.

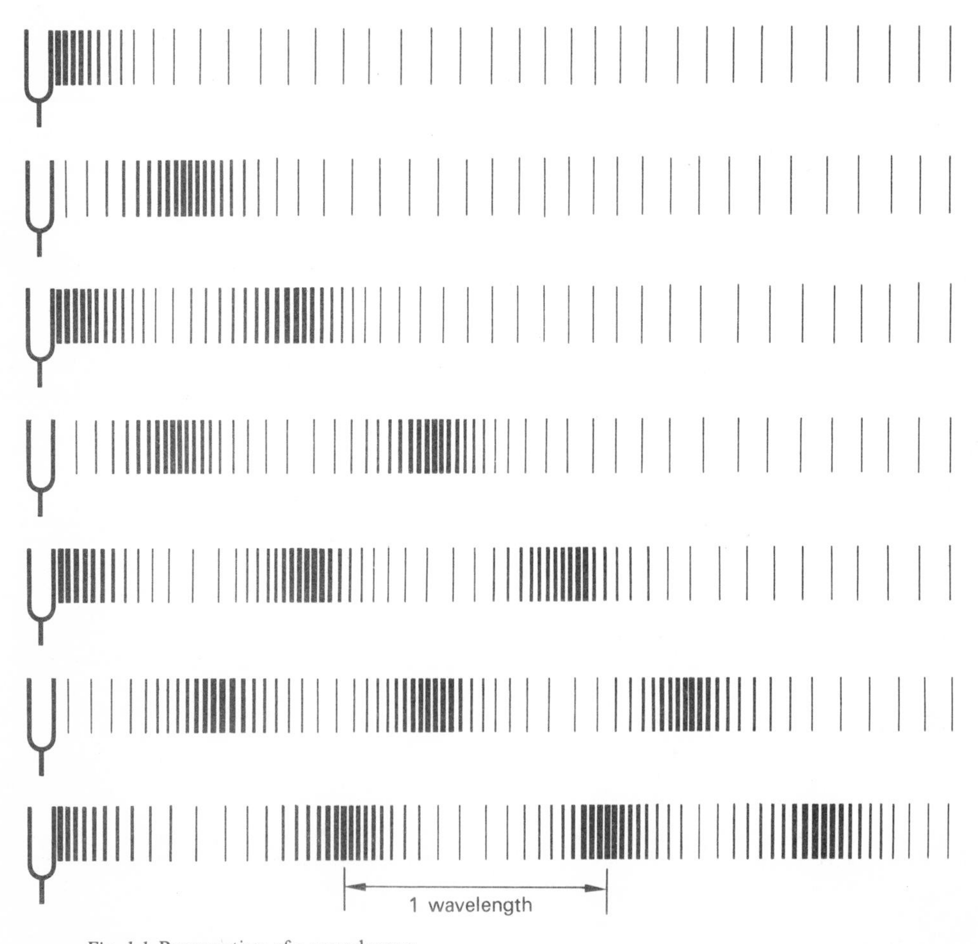


*Fig. 1.1* Propagation of a sound wave

## Simple Harmonic Motion

A pure sound consists of regular vibrations such that the displacement of the vibrating object from its original position is given by:

displacement, $s = a \sin 2\pi f \,.\, t$
where $f$ = frequency in Hz
$t$ = time in seconds from original position
$a$ = maximum displacement or amplitude

The pressure fluctuations in the air are due to molecules of air vibrating back and forth about their original position but passing on some of their energy of movement. If a particular molecule has a displacement at time $t$ of

$$s = a \sin 2\pi f \,.\, t$$

Then it is moving at a velocity of vibration given by

$$\frac{ds}{dt} = 2\pi f a \cos 2\pi f t \qquad (t = 0 \text{ when } s = 0)$$

and is being accelerated at a rate

$$\frac{d^2s}{dt^2} = -4\pi^2 f^2 a \sin 2\pi f t$$

It can be seen from Fig. 1.2 that the average displacement and pressure fluctuation is zero due to equal positive and negative changes. To overcome this problem it is convenient to make measurements of the root mean square pressure change (RMS value). For pure tones the RMS value is equal to 0·707 times the peak value or amplitude of the wave. The most commonly used measurable aspects of sound are particle displacement, particle velocity, particle acceleration and sound pressure. As the ear is a pressure sensitive mechanism it is most convenient to use pressure as the measure of sound magnitude. The sound intensity or measure of energy is related directly to the square of the sound pressure.

## Decibels

Magnitudes of sound pressure affecting the ear vary from $2 \times 10^{-5}$ N/m$^2$ at the threshold, up to 200 N/m$^2$ in the region of instantaneous damage. This may be compared with normal atmospheric pressure of $10^5$ N/m$^2$. Because of this inconveniently large order of values involved and also because the ear response is not directly proportional to pressure, a different scale is used. A psychologist, Weber, suggested that the change of subjective response (R) is proportional to the fractional change of stimulus (S) and this has been shown to be largely true.

$$\delta R \propto \frac{\delta S}{S}$$

By integration it is to be expected that the actual response will be proportional to the logarithm of the stimulus (Fechner)

$$R = k \log S$$

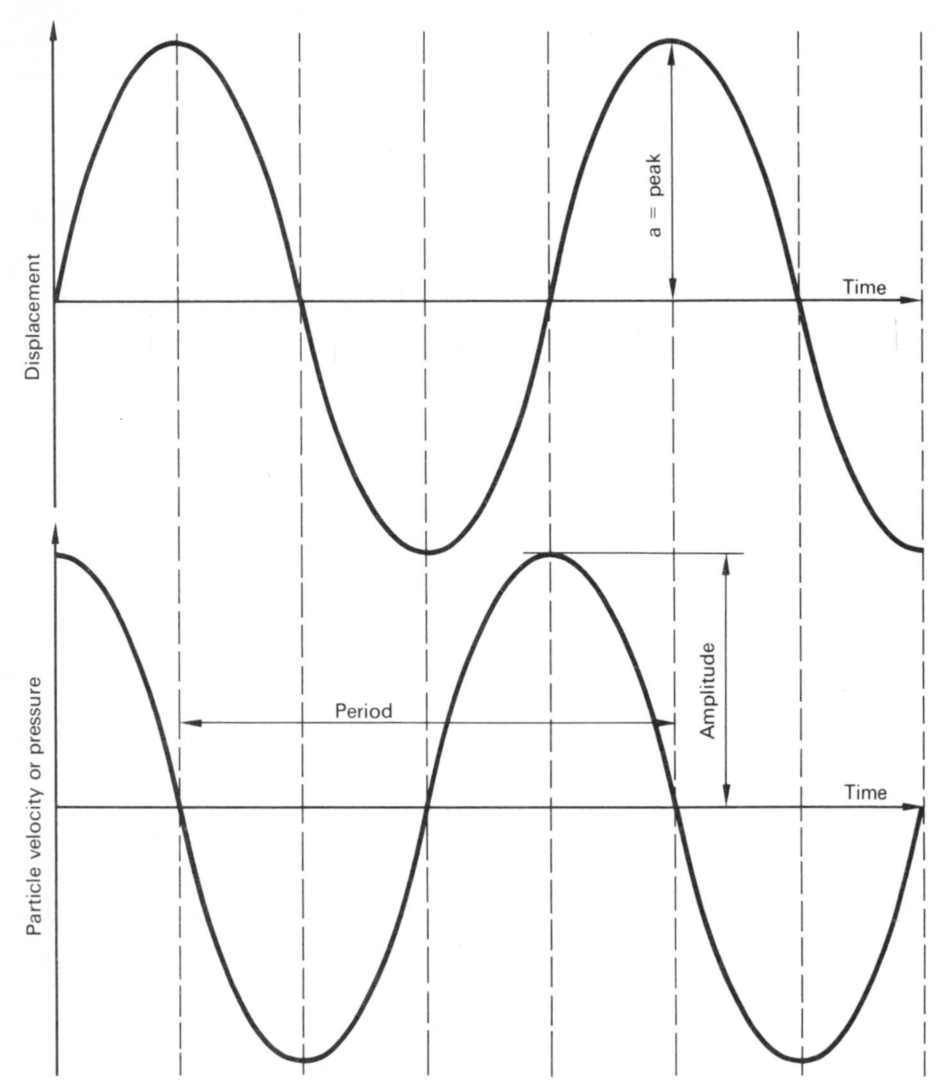


*Fig. 1.2* Displacement and pressure variations

The measure of sound pressure level used in practice, decibels, uses a logarithmic scale.

$$\text{dB} = 20 \log_{10} \frac{P_1}{P_0}$$

It will be noticed that this is not an absolute scale but a simple comparative scale relating two different pressures. For convenience $P_0$ is taken as the pressure at the threshold of hearing at 1000 Hz frequency (e.g. $2 \times 10^{-5}\ \text{N/m}^2$).

## Sound Intensity

Intensity, $I \propto P^2$

$$\therefore \quad \frac{I_1}{I_0} = \left(\frac{P_1}{P_0}\right)^2$$

$$\text{but, dB} = 10\log_{10}\left(\frac{P_1}{P_0}\right)^2$$

$$= 10\log_{10}\frac{I_1}{I_0}$$

where $I_0$ = threshold intensity, being measured in W/m$^2$.

Intensity is equal to the square of the pressure divided by the product of density times velocity of sound for the material. This product, density of air × speed of sound in air, is known as the characteristic impedance of air.

$$I = \frac{P^2}{\rho c}$$

where $\rho$ = density
$c$ = velocity of sound
$\rho c$ = 410 rayls in air

$$\therefore \quad I = \frac{P^2}{410}\text{W/m}^2.$$

$$\text{Threshold intensity} = \frac{(2\times 10^{-5})^2}{410}\text{W/m}^2$$

$$= \frac{400\times 10^{-12}}{410}$$

$$\simeq 10^{-12}\ \text{W/m}^2$$

**Example 1·1**

The RMS pressure of a sound is 200 N/m$^2$. What is the sound pressure level (S.P.L.)? (Reference pressure $2\times 10^{-5}$ N/m$^2$.)

$$\text{S.P.L.} = 20\log_{10}\frac{200}{2\times 10^{-5}}$$

$$= 20\log_{10}10^7$$
$$= 20\times 7$$
$$= 140\ \text{dB}$$

**Example 1·2**

What is the intensity of a sound whose RMS pressure is 200 N/m$^2$?

$$I = \frac{P^2}{\rho c}$$

$$= \frac{(200)^2}{410}$$

$$= \frac{40\,000}{410}$$

$$= 97{\cdot}8\ \text{W/m}^2$$

### Example 1·3

What is the sound pressure level in decibels of a sound whose intensity is 0·01 W/m$^2$?

$$\begin{aligned} \text{S.P.L.} &= 10 \log_{10} \frac{0{\cdot}01}{10^{-12}} \\ &= 10 \log_{10} 10^{10} \\ &= 10 \times 10 \\ &= 100 \text{ dB} \end{aligned}$$

### Example 1·4

What is the increase in sound pressure level (in dB) if the intensity is doubled?

$$\begin{aligned} \text{Increase in S.P.L.} &= 10 \log_{10} \frac{2I}{I_0} - 10 \log_{10} \frac{I}{I_0} \\ &= 10 \log_{10} 2 \\ &= 10 \times 0{\cdot}3010 \\ &= 3 \text{ dB} \end{aligned}$$

### Example 1·5

What is the increase in sound pressure level (in dB) if the pressure is doubled?

$$\begin{aligned} \text{Increase in S.P.L.} &= 20 \log_{10} \frac{2P}{P_0} - 20 \log_{10} \frac{P}{P_0} \\ &= 20 \log_{10} 2 \\ &= 20 \times 0{\cdot}3010 \\ &= 6 \text{ dB} \end{aligned}$$

### Addition and Subtraction of Decibels

The previous two examples illustrate that decibel values cannot be added arithmetically due to the fact that they involve logarithmic scales. Intensities can be added arithmetically but the squares of individual pressures must be added.

$$\therefore \quad I = I_1 + I_2$$

$$\text{or } P = \sqrt{(P_1^2 + P_2^2)}$$

### Example 1·6

If three identical sounds are added what is the increase in level in decibels?

$$\begin{aligned} \text{Increase in S.P.L.} &= 10 \log_{10} \frac{3I_1}{I_0} - 10 \log_{10} \frac{I_1}{I_0} \\ &= 10 \log_{10} 3 \\ &= 10 \times 0{\cdot}4771 \\ &= 4{\cdot}8 \text{ dB} \\ &\simeq 5 \text{ dB} \end{aligned}$$

When adding decibel values it is necessary to use a logarithmic scale and it is convenient for this purpose to use a chart such as Fig. 1.3.

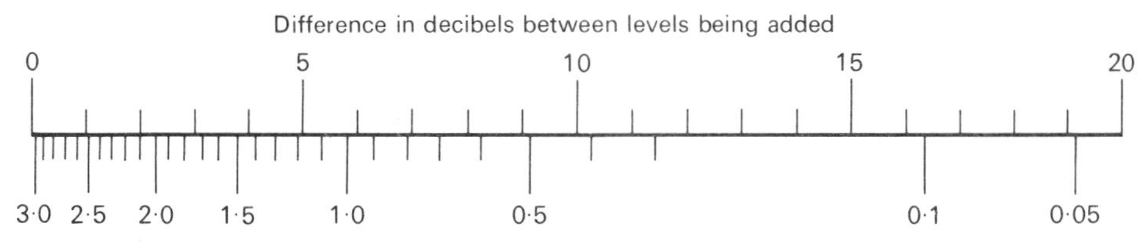


*Fig. 1.3* Scale for combining sound pressure levels

### Example 1·7

Two cars are producing individual sound pressure levels of 77 dB and 80 dB measured at the pavement. What is the resultant sound pressure level when they pass each other?

Difference = 80 − 77
= 3 dB

From Fig. 1.3 amount to be added to the higher level
= 1·75 dB

Resultant S.P.L. = 81·75 dB

### Example 1·8

In a certain factory space the noise level with all machines running is 101 dB. One machine alone produces a level of 99 dB. What would the level be in the factory with all except this machine running?

Difference in sound pressure level
= 101 − 99
= 2 dB

amount to be subtracted from the higher level from Fig. 1.3
= 2·4 dB

Resultant sound pressure level
= 96·6 dB

### Sound Level Meters and Weighting Scales

The sound level meter used for the measurement of RMS sound pressure levels consists of a microphone, amplifier and a meter. The microphone converts the sound pressure waves into electrical voltage fluctuations which are amplified and operate the meter. Unfortunately no meter could indicate accurately over such a large range as may be needed from 30 dB to 120 dB or more. To overcome this the amplification is altered as required in steps of 10 dB and the meter only has to read the difference between the amplifier setting and the sound pressure level. Most meters will have connections to which filters can be added to select particular frequencies of the sound. An output is common to allow the sound to be recorded on tape or the levels plotted on a chart.

Besides a linear reading of sound pressure level most meters have A and B scales where the response varies with frequency as shown in Fig. 1.4. It can be seen that a 30 Hz note of sound pressure level 70 dB would indicate 70 − 40 = 30 dBA, or 70 − 17 = 53 dBB or 70 − 3 = 67 dBC. The C scale is taken to be linear for most

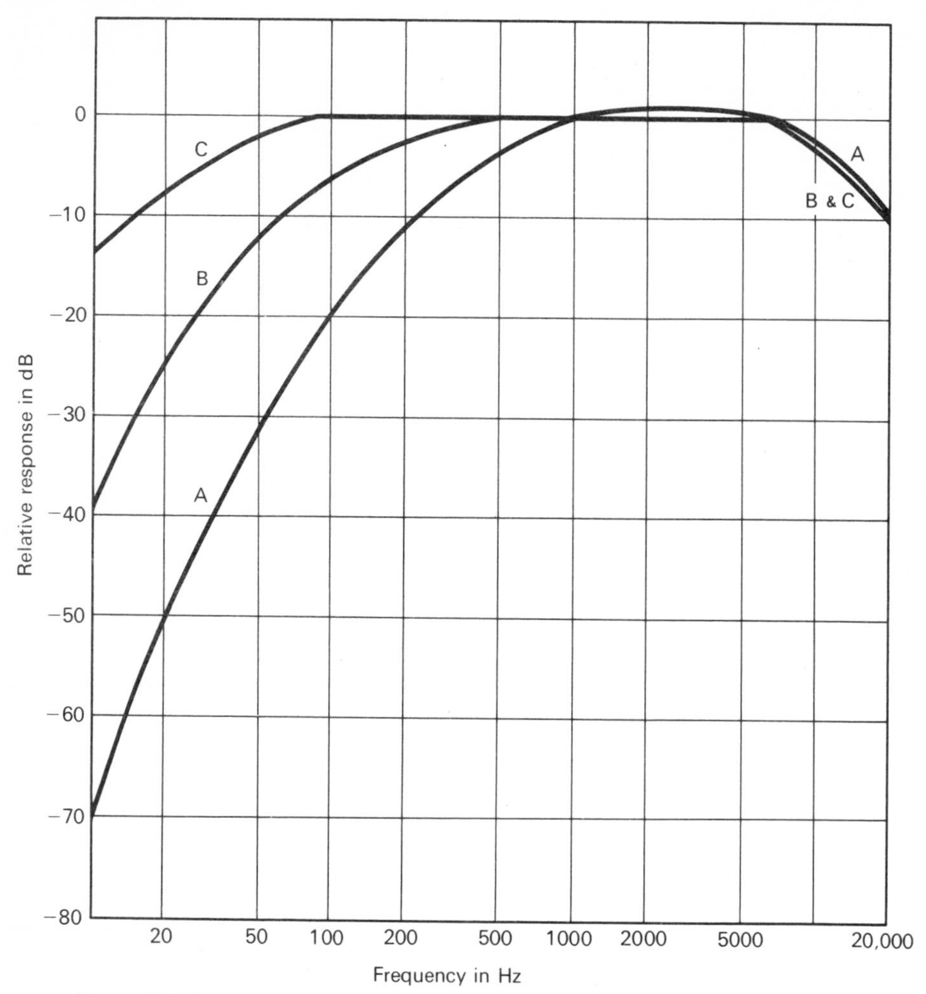


*Fig. 1.4* Relative response of A, B and C weighting scales

practical purposes, but in fact is so only for frequencies from 200 to 1250 Hz. The attenuation either side is small. The B scale is intended to give responses on the sound level meter corresponding to the 70 dB equal loudness contour for pure tones (see Chapter 2). The A response corresponds approximately to the 40 dB equal loudness contour.

It has been shown that the readings on the A scale, dBA, correspond most closely to the response of the ear. For many practical purposes when simple direct readings are needed this is the best scale to use. It is shown in the next chapter that the response of the ear is dependent on frequency, and readings of sound pressure on a linear scale can be most misleading in subjective acoustics. It was for this reason that the weighting scales were originally devised.

## Calibration of Sound Level Meters

This is done by means of a source of known noise level, such as a falling ball calibrator or pistonphone (Figs. 1.5 and 1.6).

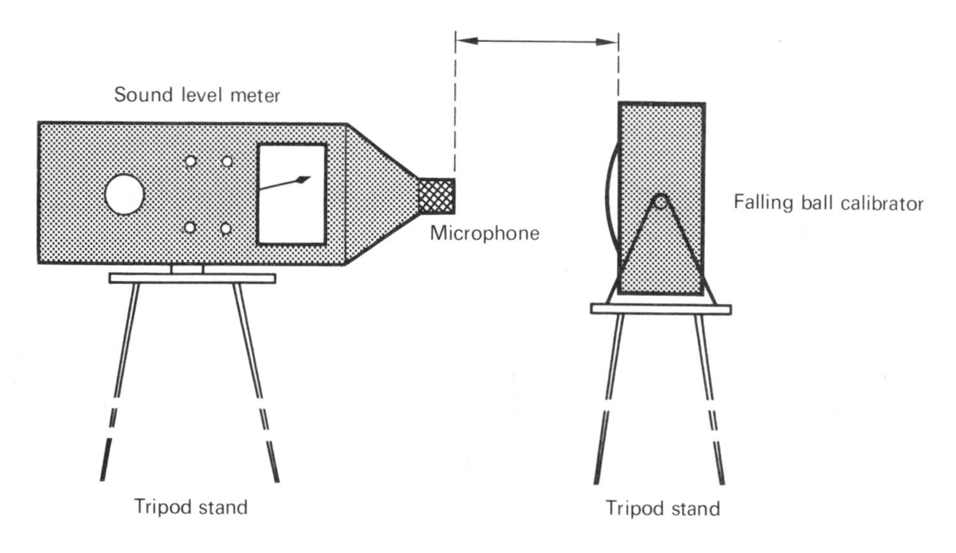


*Fig. 1.5* Calibration of a sound level meter using a falling ball

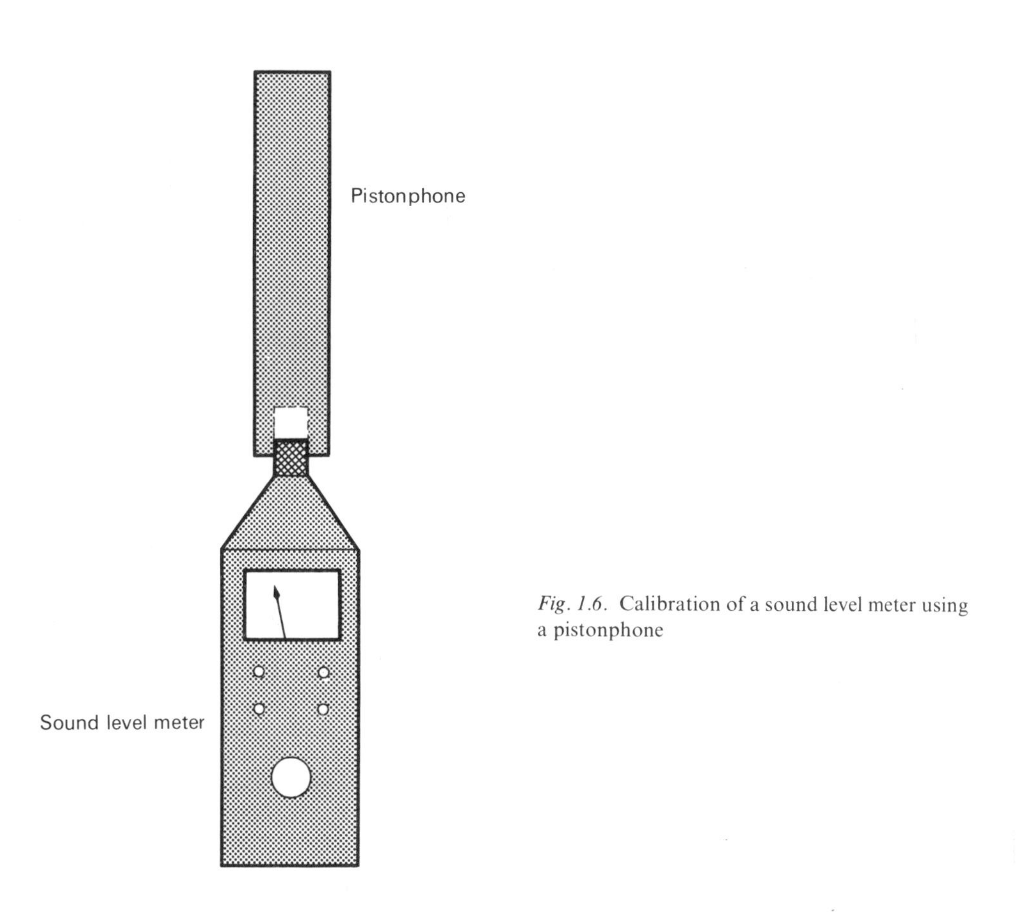


*Fig. 1.6.* Calibration of a sound level meter using a pistonphone

The falling ball calibrator consists of a container where a large number of steel balls fall onto a diaphragm. This produces a broad band of sound of about 90 dB at 100 mm distance. Meters may be checked and slight adjustments made if necessary. It is vital that the correct conditions such as distance are obtained according to the maker's instructions.

Calibration by the pistonphone is slightly simpler in that there is no difficulty in positioning the source and meter. The pistonphone fits over the microphone and produces a note of 250 Hz at 124 dB. Exterior noise is unimportant, but the calibrator can only be used on microphones to which it will fit. A small motor operates a cam, producing the sound by sinusoidal movement of two small pistons. This second method is also very convenient as a standard source of sound when making calibrated tape recordings for later analysis in the laboratory.

## Analysis of Sound

Sound does not normally consist of single frequency notes, but a highly complex combination of tones. Often, it is necessary to know at least what bands of frequencies are present. Mostly it is sufficient to know the magnitude of the sounds contained within octave bands: 75–150 Hz, 150–300 Hz, 300–600 Hz, 600–1200 Hz, 1200–2400 Hz, 2400–4800 Hz, 4800–9600 Hz. It can be seen that one octave band consists of all sounds from any frequency to twice that same frequency. Similarly for third octave bandwidths. In each case it is convenient to refer only to the centre frequency within each band; e.g. 63, 125, 250, 500, 1000, 2000, 4000, 8000 Hz for octaves and 63, 100, 125, 160, 200, 250, 315, 400, etc. for third octaves.

## Filters

Using suitable electrical filters, it is possible to separate the appropriate band of frequencies from the remainder, thus measuring the magnitude of that one group. In practice, filters working to one octave or third octave bandwidths are most common for building acoustics. Even third octave analysis will involve three times as much work and time as octave analysis. Narrower bandwidth filters down to one or two cycles width are available but they are of very limited application because of the amount of time involved in their use. The centre frequencies of octave, half octave and third octave bandwidths used in standard filters are given in Table 1·1.

**Table 1·1**

*Preferred Frequencies for Acoustic Measurements*

| |
|---|
| *Octave*<br>16, 31·5, 63, 125, 250, 500, 1000, 2000, 4000, 8000 Hz |
| *Half Octave*<br>16, 22·4, 31·5, 45, 63, 90, 125, 180, 250, 355, 500, 710, 1000, 1400, 2000, 2800, 4000, 5600, 8000 Hz |
| *Third Octave*<br>16, 20, 25, 31·5, 40, 50, 63, 80, 100, 125, 160, 200, 250, 315, 400, 500, 630, 800, 1000, 1250, 1600, 2000, 2500, 3150, 4000, 5000, 6300, 8000 Hz |

These are the geometric centre frequencies of filter pass bands.

### Example 1·9

A certain noise was analysed into octave bands. The sound pressure levels in each were measured as shown below. What was the total level?

| Centre Frequency Hz | 125 | 250 | 500 | 1000 | 2000 | 4000 |
|---|---|---|---|---|---|---|
| S.P.L. in dB | 80 | 82·5 | 77·5 | 70 | 65 | 60 |

*Method 1*

Total intensity $I = I_1 + I_2 + I_3 + \ldots I_6$

$$\text{But dB} = 10 \log \frac{I_1}{I_0}$$

| Frequency Hz | S.P.L. in dB | $\frac{I}{I_0} \times 10^8$ |
|---|---|---|
| 125 | 80 | 1·0000 |
| 250 | 82·5 | 1·7780 |
| 500 | 77·5 | 0·5623 |
| 1000 | 70 | 0·1000 |
| 2000 | 65 | 0·0316 |
| 4000 | 60 | 0·0100 |
| | | 3·4819 |

$$\begin{aligned} \text{Total Sound Pressure Level} &= 10 \log_{10} 3{\cdot}4819 \times 10^8 \\ &= 10 \times 8{\cdot}5419 \\ &= 85{\cdot}5 \text{ dB} \end{aligned}$$

*Method 2*

The levels may be combined in pairs using Fig. 1.3.

| Frequency Hz | S.P.L. in dB | Difference in dB | Add dB | Result | Add dB | Result | Result |
|---|---|---|---|---|---|---|---|
| 125 | 80 | | | | | | |
| | | 2·5 | 1·95 | 84·5 | | | |
| 250 | 82·5 | | | | | | |
| | | | | | 0·95 | 85·5 | |
| 500 | 77·5 | | | | | | |
| | | 7·5 | 0·75 | 78·25 | | | 85·5 |
| 1000 | 70 | | | | | | |
| 2000 | 65 | | | | | | |
| | | 5·0 | 1·2 | 67·2 | | 67·2 | |
| 4000 | 60 | | | | | | |

### Example 1·10

Calculate the sound pressure level in dBA of a noise with the following analysis:

| Centre Frequency Hz | 31·5 | 63 | 125 | 250 | 500 | 1000 | 2000 | 4000 |
|---|---|---|---|---|---|---|---|---|
| S.P.L. in dB | 60 | 60 | 65 | 70 | 65 | 65 | 45 | 40 |

From Fig. 1.4 the dBA values in each octave are found, then added as shown in the previous example.

| Frequency Hz | Level dBA | Sum dBA | Sum dBA | Sum dBA |
|---|---|---|---|---|
| 31·5 | 20 | | | |
| | | 33·2 | | |
| 63 | 33 | | | |
| | | | 62·2 | |
| 125 | 48 | | | |
| | | 62·2 | | |
| 250 | 62 | | | |
| | | | | 68·1 |
| 500 | 62 | | | |
| | | 66·8 | | |
| 1000 | 65 | | | |
| | | | 66·8 | |
| 2000 | 45 | | | |
| | | 46·2 | | |
| 4000 | 40 | | | |

Total level in dBA = 68·1.

### Sound Power

Total sound power in watts is equal to the intensity in watts/$m^2$ multiplied by the area in $m^2$. A large orchestra may produce 10 watts, a jet plane up to 100 kW, whereas the loudest voice would only reach about 1 mW.

It is convenient to put the sound power level on a logarithmic scale relative to a standard power level. The standard is $10^{-12}$ watts.

$$\text{Sound power level, PWL} = 10 \log_{10} \frac{W}{W_0}$$

$$= 10 \log_{10} \frac{W}{10^{-12}}$$

$$= 10 \log_{10} W - 10 \log_{10} 10^{-12}$$

$$= 10 \log_{10} W + 120$$

**Example 1·11**

Determine the sound power level of 0·001 watts.

$$\begin{aligned}\text{PWL} &= 10\log_{10} 0{\cdot}001 + 120\\ &= 10\log_{10} 10^{-3} + 120\\ &= -30 + 120\\ &= 90\ \text{dB re } 10^{-12} W\end{aligned}$$

It will have been noticed that both sound power level and sound pressure level are given in dB. Whenever giving sound power levels it is essential to state the reference level.

### Sound Power and Sound Intensity

The sound intensity from a point source of sound radiating uniformly into free space can be found from the power output and the distance from the source, $r$.

$$\text{Intensity, } I = \frac{\text{Sound power in watts}}{4\pi r^2}$$

If the sound is produced at ground level, assuming that the ground is perfectly reflecting, then the energy is only radiated into a hemisphere instead of a complete sphere. In this case the formula for intensity becomes:

$$I = \frac{W}{2\pi r^2}$$

**Example 1·12**

Calculate the intensity and S.P.L. of a sound at a distance of 10 m from a uniformly radiating source of 1 watt power.

$$\begin{aligned}I &= \frac{W}{4\pi r^2}\\ &= \frac{1{\cdot}0}{4\pi(10)^2}\\ &= 7{\cdot}95 \times 10^{-4}\ \text{W/m}^2\end{aligned}$$

$$\begin{aligned}\text{S.P.L.} &= 10\log_{10}\frac{7{\cdot}95\times 10^{-4}}{10^{-12}}\\ &= 10\log_{10} 7{\cdot}95 \times 10^{8}\\ &= 10 \times 8{\cdot}9004\\ &= 89\ \text{dB}\end{aligned}$$

**Example 1·13**

Find the power output of a sound which radiates uniformly into unobstructed space if the pressure at a distance of 5 m is 1 $\text{N/m}^2$. (Assume $\rho c = 400$ rayls)

$$\text{Intensity, } I = \frac{p^2}{\rho c}$$

$$\therefore \quad \frac{P^2}{\rho c} = \frac{W}{4\pi r^2}$$

$$\therefore \quad W = \frac{4\pi r^2 P^2}{\rho c}$$

$$= \frac{4\pi 25 \cdot 1^2}{400}$$

$$= 0{\cdot}786 \text{ watts}$$

## Questions

(1) Calculate the sound pressure level in dB of a sound whose root mean squared pressure is 7·2 N/m$^2$.

(2) Determine the sound pressure level in dB of a sound whose intensity is 0·007 W/m$^2$ (re $10^{-12}$ W/m$^2$).

(3) The noise level from a factory with ten identical machines measured near some residential property was found to be 54 dB. The maximum permitted is 50 dB at night. How many machines could be used during the night?

(4) Find the total sound pressure level in dB for a sound with the following analysis. Calculate also the total intensity in W/m$^2$.

| Centre Frequency Hz | Level dB |
|---|---|
| 125 | 55 |
| 250 | 63 |
| 500 | 71 |
| 1000 | 68 |
| 2000 | 59 |

(5) A motor car was found to produce the following noise. Calculate the total noise level in dB (linear) and dB (A).

| Octave Band Hz | Level dB |
|---|---|
| 20–75 | 95 |
| 75–150 | 84 |
| 150–300 | 80 |
| 300–600 | 68 |
| 600–1200 | 65 |
| 1200–2400 | 61 |
| 2400–4800 | 60 |
| 4800–10 000 | 60 |

(6) Calculate the intensity of a sound whose RMS pressure is 0·0045 $N/m^2$ if $\rho c$ = 410 rayls for air.

(7) Find the RMS pressure of a sound whose intensity is 1 $W/m^2$.

(8) Determine the sound power level range (re $10^{-12}$ W) of the human voice which is from about 10 to 50 microwatts.

(9) Calculate the intensity and S.P.L. of a sound at a distance of 20 m from a uniformly radiating source of 1·26 W power.

(10) Two sounds of 4 W and 10 W power are produced at ground level at a distance of 10 m and 20 m respectively from a listener. If the ground is level, unobstructed and non-absorbing, what will be the S.P.L. of the sound heard by the listener?

(11) What power output is needed to produce a S.P.L. of 60 dB at a distance of 100 m if both source and hearer are at ground level? Assume that the ground is level, unobstructed and non-absorbing.

(12) Assuming that it radiates sound uniformly, find the sound power from a motor car whose S.P.L. at a distance of 7·5 m is 87 dB.

Chapter 2

# Aural Environment

In the first chapter the physical properties of sounds and their measurement were discussed. In this chapter the subjective effects of noise and their cause are examined.

## The Ear

The ear is a transducer converting sound pressure waves into signals which are sent to the brain (see Figs. 2.1 and 2.2).

Sound first reaches the outer and visible part of the ear known as the concha. A concave shape of a certain size will act as a focusing device only for wavelengths up to the same order of size, and so the concha will tend to scatter the longer wavelengths whilst reflecting shorter ones into the meatus. The meatus is the tube connecting the outer ear to the ear drum, and because of its size, it resonates to a frequency of about 3 kHz. The ear drum separates the outer from the inner ear. Major atmospheric pressure changes can be equalized on either side of the ear drum through the eustachian tube by the act of swallowing. The problem at this stage is a high impedance mismatch due to the outer and middle ear being filled with air and the inner ear filled with liquid. The small bones which connect the ear drum with the oval window,

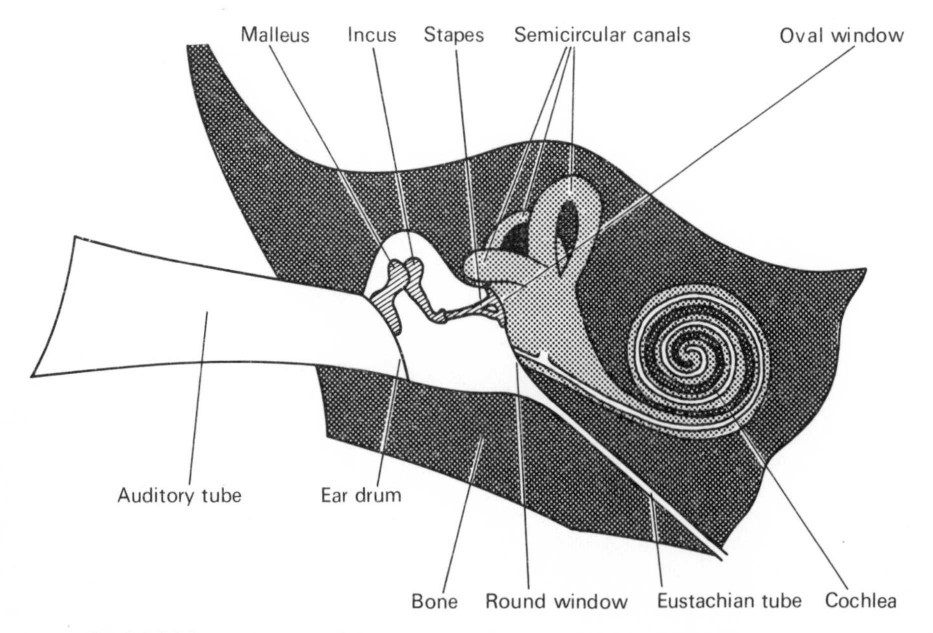


*Fig. 2.1* Main components of the ear, showing outer, middle and inner parts

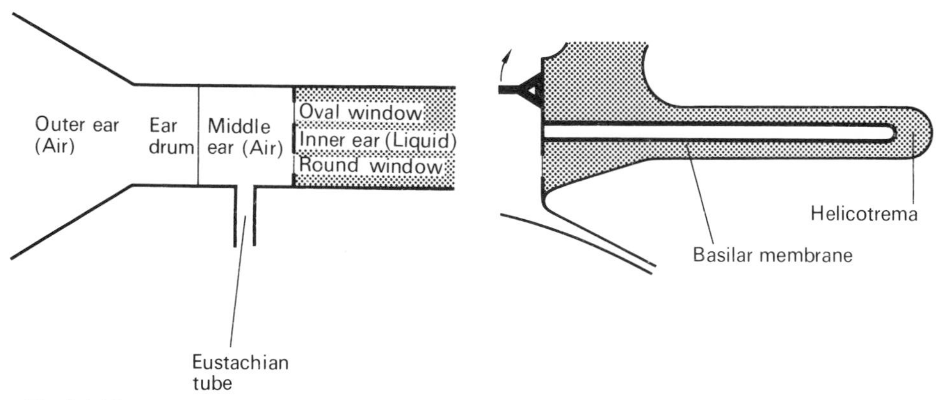


*Fig. 2.2* Three main sections of the ear shown schematically

*Fig. 2.3* Diagram of the cochlea unwound to show its main components

effectively make the middle ear equivalent to a 'step up transformer' of about twenty times. This just compensates for a theoretical loss of approximately 30 dB between air and fluids of the inner ear.

The lever system consists of three bones when in theory only one is needed (as is the case with birds). It does give three important advantages:

1. Minimum bone conduction.
2. More linear response for different frequencies.
3. A protective overload device possible as the ossicles change their mode of operation above 140 dB sound pressure level.

The middle ear also possesses another protective device consisting of two small muscles which adjust the ear drum and stapes for levels of sound above 90 dB which last more than 10 ms.

The inner ear is a system of liquid filled canals protected both mechanically and acoustically by being located inside the temporal bone of the skull. The cochlea is a hollow coil of bone filled with liquid, with a total length of about 40 mm (see Fig. 2.3). This is divided along its length by the basilar membrane with a small gap at its far end known as the helicotrema. Acoustic energy is converted into impulses transmitted to the brain at the basilar membrane on which about 24 000 nerve endings terminate.

Intense sounds can damage or even destroy any of the moving parts of the ear. In the more common case of hearing damage, because of prolonged exposure to high levels of noise, it is the hair cells that are damaged.

## Audible Range

This depends upon the age and physical condition of a person, but can be from 20 Hz to 20 kHz. The sensitivity varies considerably over this frequency range especially near the threshold of hearing where there is a variation of the order of 70 dB, as shown in Fig. 2.4.

## Pitch

Frequency is an objective measure of the number of vibrations per second, whereas the term pitch is subjective, and although dependent mainly on frequency is also affected by intensity.

## Loudness

The loudness of a sound is a subjective effect which is a function of the ear and brain as well as amplitude and frequency of the vibration. In practice it is usual to consider people with normal hearing and correlate only amplitude and frequency with loudness. Pure tones of different frequencies are compared with that of 1000 Hz by adjusting the amplitude to obtain equal loudness contours. The loudness level is given in phons. These are numerically equal to the sound pressure level in dB at 1 kHz as shown in Fig. 2.5.

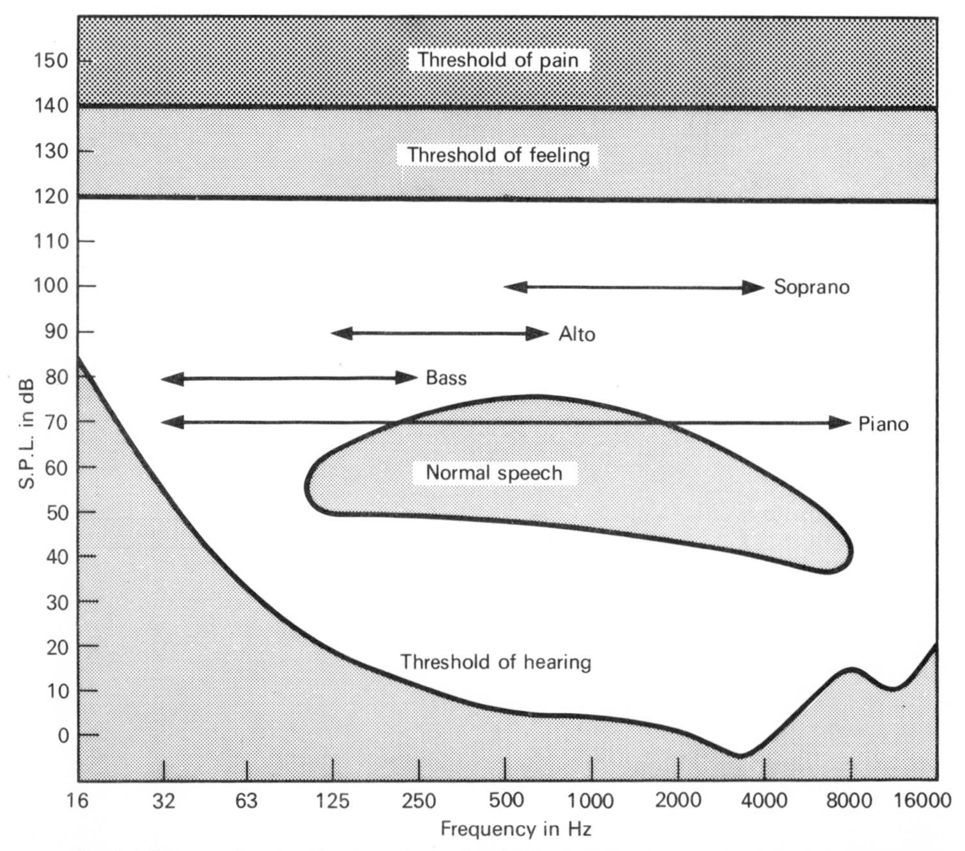


*Fig. 2.4* Diagram showing the approximate threshold of hearing for young people in the age range 18–25. The thresholds of feeling and pain occur at about 120 and 140 dB respectively. The range of levels and frequencies of normal speech are shown. Frequency limits for the piano and singers are shown

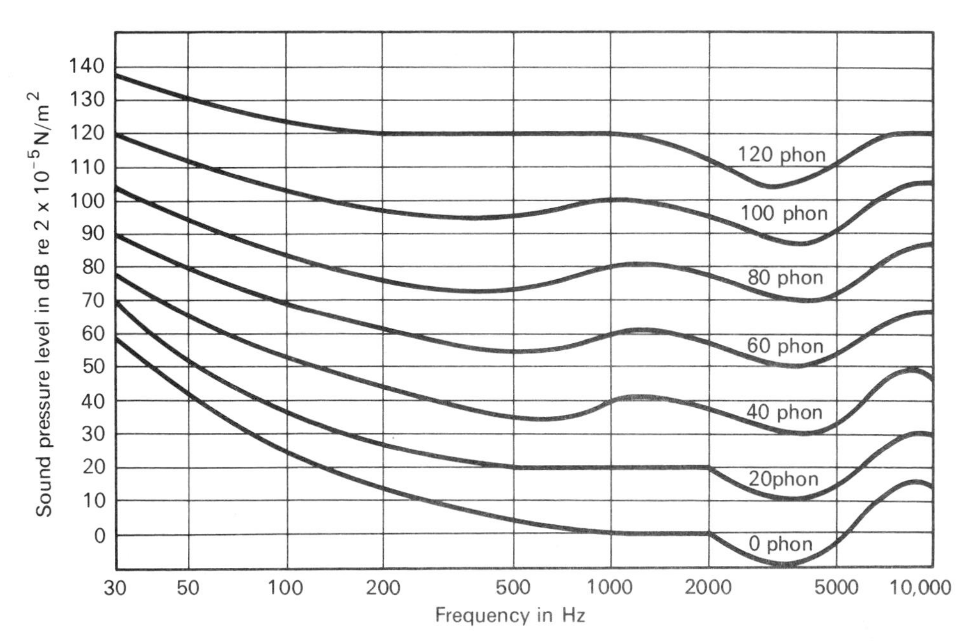


*Fig. 2.5* Equals loudness contours

The phon scale is quite arbitrary and has no physical or physiological basis, but is convenient because of its relation to decibels (at 1 kHz). Unfortunately it is not conveniently additive for different sounds and another scale, that of sones, which overcomes this difficulty is used (See Fig. 2.6). The values of sones may be added arithmetically to obtain loudness levels, e.g. 50 sones is twice as loud as 25 sones.

It is obviously very easy to find the loudness in phons or sones of pure tones from Fig. 2.5. When the noise is other than pure it is commonly assessed in terms of Stevens phons. An octave analysis (or third octave) of the noise needs to be done in the eight bands 20–75, 75–150, 150–300, 300–600, 600–1200, 1200–2400, 2400–4800 and 4800–10 000 Hz. Each level in dB is converted to sones by means of Fig. 2.7. Then the total loudness in sones, $S_t$ is given by:

$$S_t = S_m + F(\Sigma S - S_m)$$

where $S_m$ = loudness of loudest band

and $F$ = 0·15 for third octave bandwidth

= 0·2 for half octave bandwidth

= 0·3 for octave bandwidth

and $\Sigma S = S_1 + S_2 + S_3 + S_4 + S_5 + S_6 + S_7 + S_8$

($S_1$, $S_2$, etc. = loudness in appropriate bands)

This total level in sones, $S_t$, may be then converted into phons by means of Fig. 2.6.

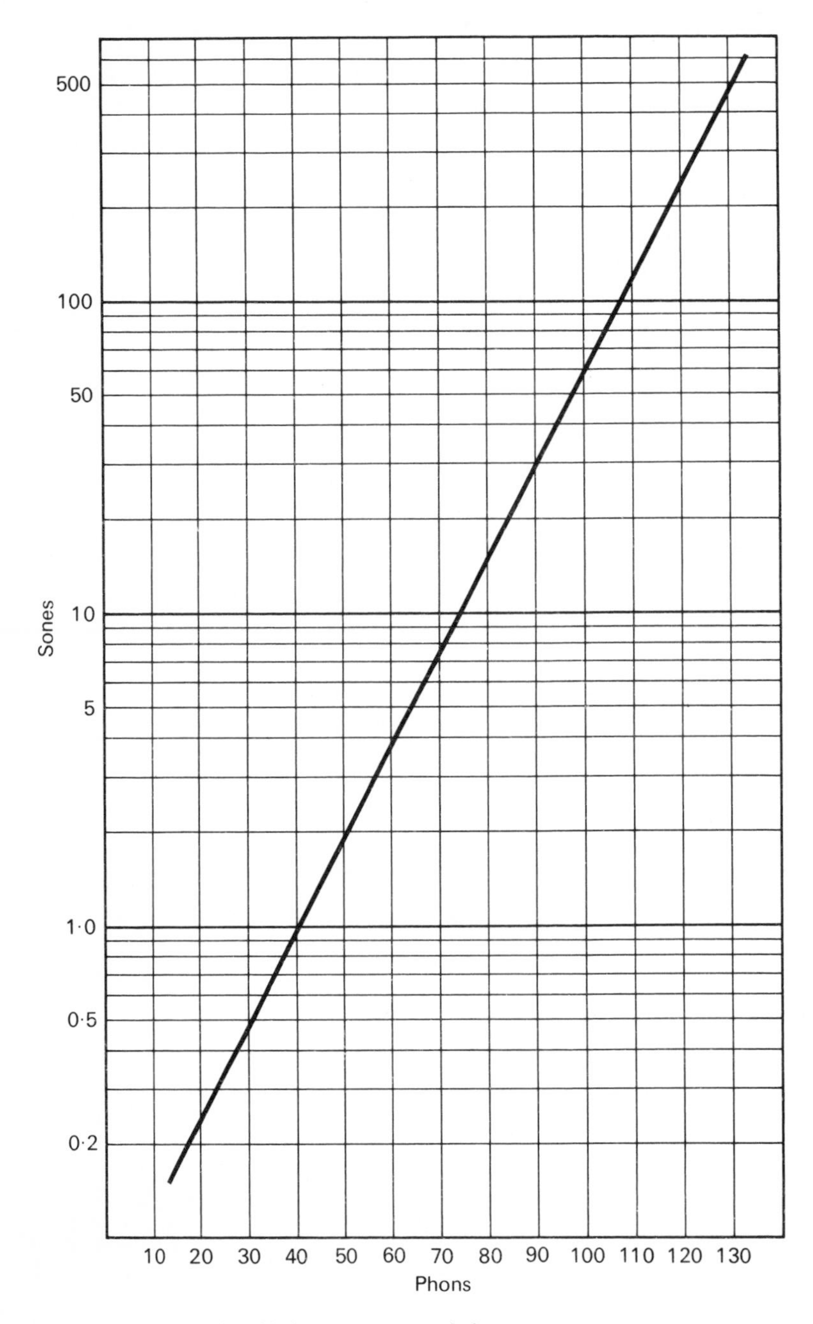


*Fig. 2.6* Relationship between sones and phons

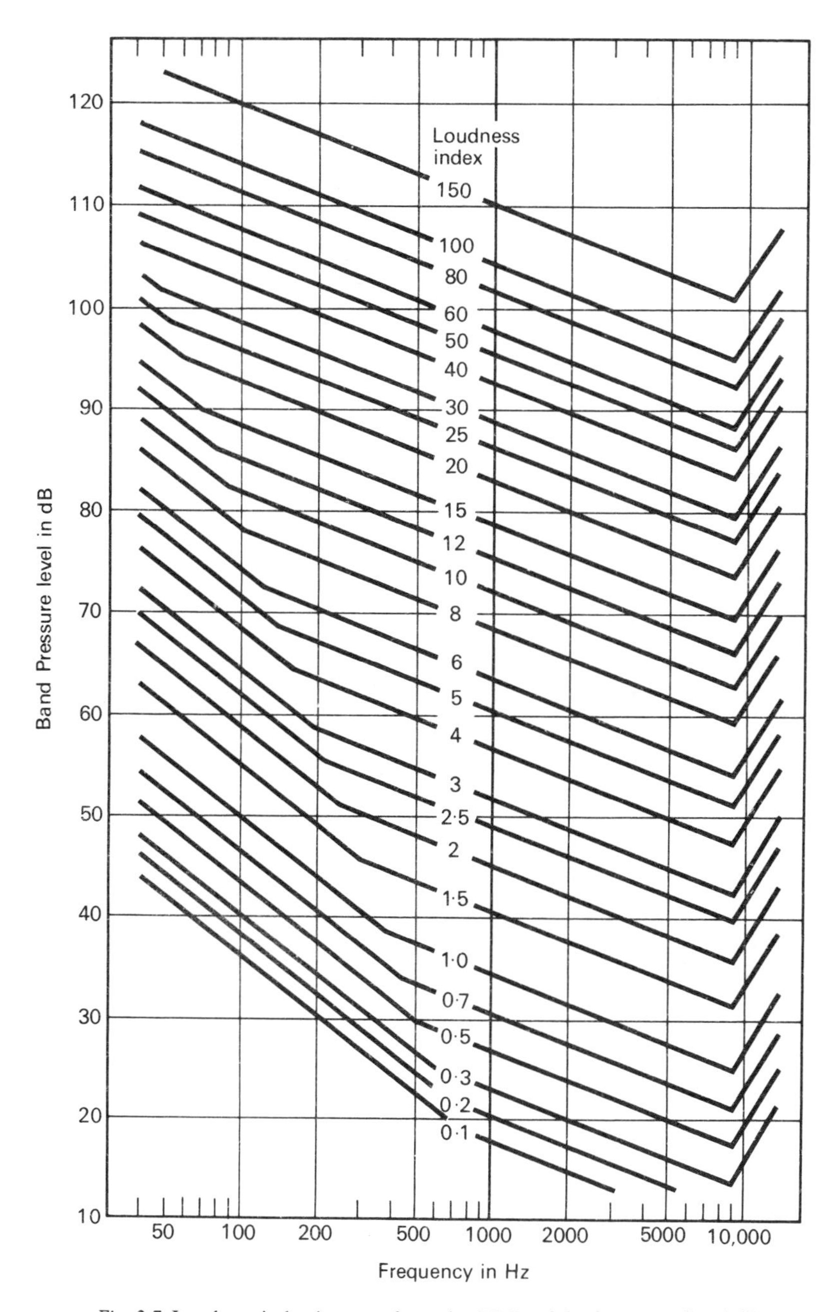


*Fig. 2.7* Loudness index in sones from the S.P.L. of the frequency band dB

### Example 2·1

A certain motor car was found to produce the following noise levels:

| *Octave band* | *dB level* |
|---|---|
| 20–75 | 95 |
| 75–150 | 84 |
| 150–300 | 80 |
| 300–600 | 68 |
| 600–1200 | 65 |
| 1200–2400 | 61 |
| 2400–4800 | 60 |
| 4800–10 000 | 60 |

Calculate the level in phons.

Using Fig. 2.7, convert the dB readings into sones.

| *Octave band* | *dB level* | *Sones* |
|---|---|---|
| 20–75 | 95 | 20 |
| 75–150 | 84 | 12 |
| 150–300 | 80 | 11 |
| 300–600 | 68 | 7 |
| 600–1200 | 65 | 6 |
| 1200–2400 | 61 | 6·5 |
| 2400–4800 | 60 | 7 |
| 4800–10 000 | 60 | 8 |

$$\Sigma S = 77{\cdot}5$$

$$S_m \text{ (the loudest)} = 20 \text{ sones}$$

$$S_t = 20+0{\cdot}3(77{\cdot}5-20)$$
$$= 20+0{\cdot}3(57{\cdot}5)$$
$$= 37.25 \text{ sones}$$

From Fig. 2.6 the level in phons = *92*

### Table 2·1

*Loudness of Common Noises*

| Noise | Sones | Phons |
|---|---|---|
| Large jet plane 80 m overhead | 700 | 134 |
| Heavy road traffic at kerbside | 79 | 103 |
| Light road traffic at kerbside | 16 | 80 |
| Students' refectory | 20 | 84 |
| Normal speech (male) at 1 m | 11 | 75 |
| Inside noisy motor car | 40 | 94 |
| Machine shop | 97 | 106 |

### Octave Band Analysis

In the last section the need was seen for an octave analysis in order to calculate loudness. (It could have been done using a third octave analysis but this is long and tedious and seldom necessary.)

There are two ways to measure the levels in each octave band. The first is by the use of a sound level meter set on to the linear scale using a set of octave band filters. For noises which are reasonably continuous this is simple. Where the noise is more conveniently recorded and analysed in the laboratory a calibrated recording must be made (see Fig. 2.8). The output of the sound level meter is connected to a portable tape recorder. A noise of known level must be recorded on the tape. This is normally done by means of a pistonphone or falling ball calibrator. This signal, attenuated by a known amount, using the sound level meter is fed to the tape recorder. The actual noise is recorded with a suitable attenuation on the sound level meter. The whole is later played back through an octave filter set (or audio frequency analyser which operates on the same principle) and the levels recorded on paper (see Fig. 2.9). A very short piece of tape can be played repeatedly for each octave band.

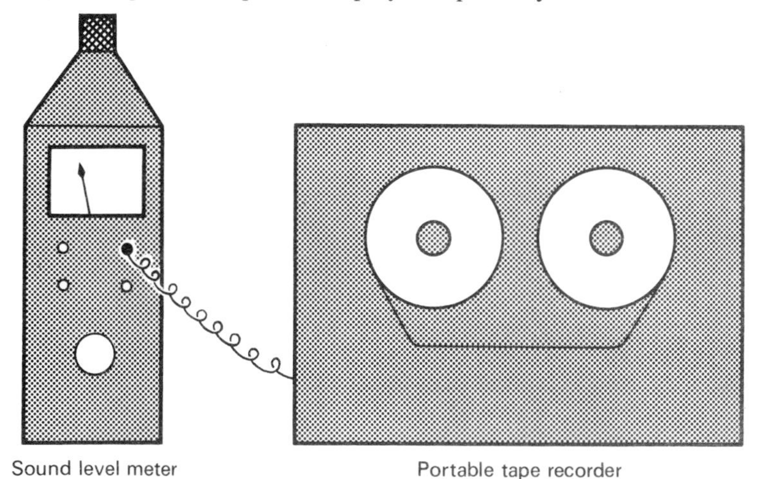


*Fig. 2.8* Calibrated recording of sound on tape using a sound level meter to give a known attenuation. The calibrated signal is put on the tape using a pistonphone as sound source

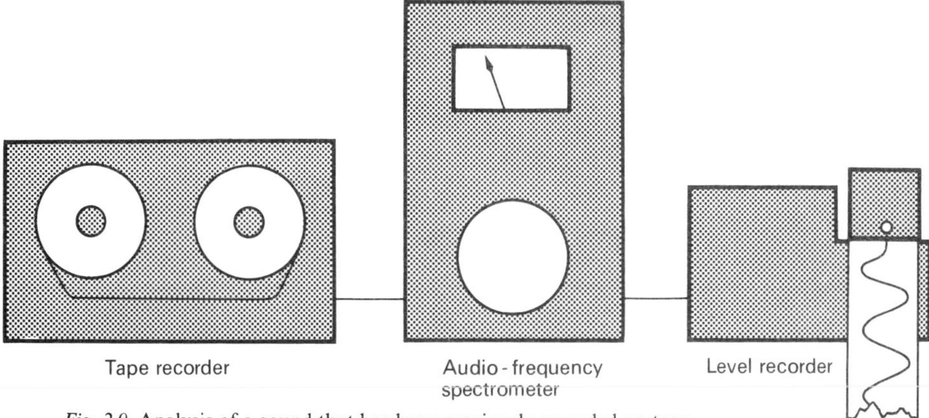


*Fig. 2.9* Analysis of a sound that has been previously recorded on tape

**Table 2·2**

*Typical Octave Band Analysis Levels in dB*

| Noise | Centre frequency of octave band, Hz | | | | | | | |
|---|---|---|---|---|---|---|---|---|
| | 63 | 125 | 250 | 500 | 1000 | 2000 | 4000 | 8000 |
| Light traffic at kerbside | 81 | 81 | 75 | 70 | 72 | 71 | 63 | 60 |
| Heavy traffic at kerbside | 96 | 93 | 90 | 88 | 89 | 84 | 78 | 76 |
| Machine shop | 68 | 72 | 90 | 87 | 86 | 88 | 90 | 84 |
| Students' refectory | 68 | 70 | 75 | 75 | 68 | 64 | 56 | 49 |

### Masking

This is the effect of one noise on another. Speech, for instance, can become masked by road traffic or aircraft noise. The masking noise raises the threshold of audibility for other sounds. A masking noise is most effective with a sound of similar frequency, but a low frequency noise is more effective in masking one of high frequency than the reverse.

Masking can obviously be inconvenient where very good hearing conditions are needed, but it can also be useful when speech privacy is required. For instance, it may be utilized by playing soft music in the waiting room of a doctor's surgery to prevent those waiting hearing details of examinations. A better solution would be good sound insulation. It thus follows that the amount of sound insulation needed depends upon the ambient noise level, which can provide a useful masking effect. Open plan offices have been designed with reasonable audible privacy on the basis of the admission of a certain amount of exterior traffic noise. Although the calculation of the amount by which a masking noise raises the threshold of audibility is complicated, it should be noted that even relatively low levels of background noise have a considerable masking effect.

### Audiometry

Audiometry is the term used to describe the measurement of hearing sensitivity. The instrument used for this purpose is an audiometer which produces pure tones of various frequencies at different known pressure levels. The subject is asked to say, for each ear, which level at each frequency he can just detect. With no hearing defect the audiometer results follow the bottom curve of Fig. 2.4. If the subject has some hearing defect, the audiometer levels would need to be raised at some or all frequencies above the normal threshold. The amount the level needs to be increased gives the hearing loss. Examples of audiometric tests for left and right ears are given in Fig. 2.10.

### Hearing Defects

These are usually divided into two general types. First, those which are not related to noise exposure; and second, those which are directly attributable to damage caused by noise.

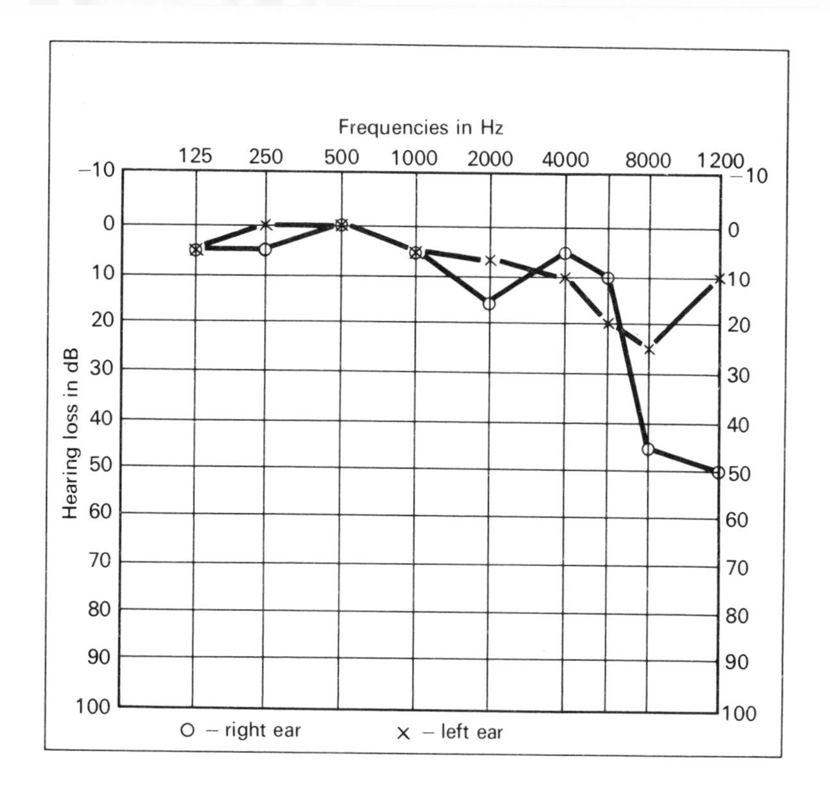


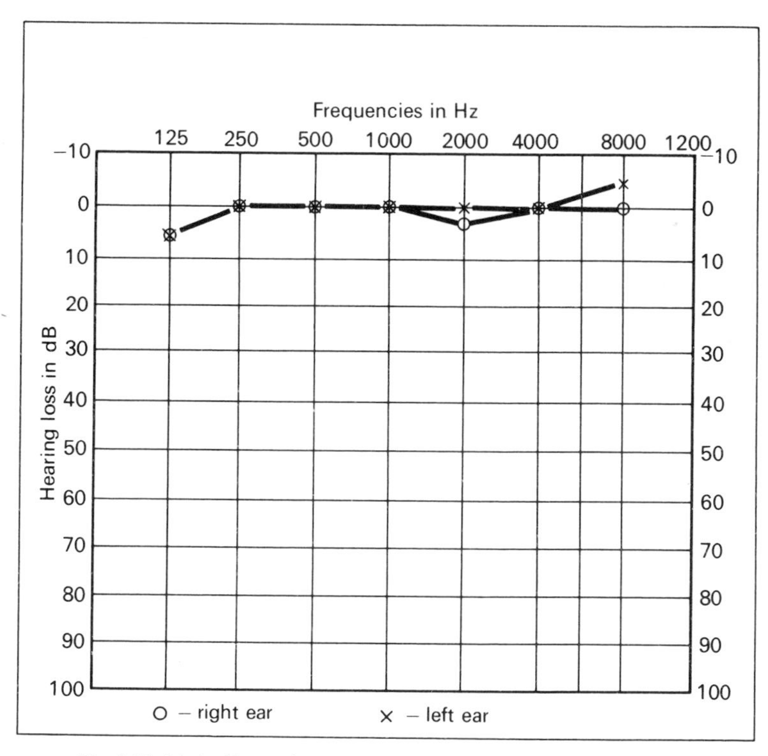


*Fig. 2.10* (*a*) Audiometric test result showing some hearing loss at high frequencies
(*b*) Audiometric test result showing normal hearing

(1) Defects Not Caused by Noise

*Presbycousis*
This is a loss which is normally associated with age. Whether this is simply due to age or whether it is due to the effects of normal levels of environmental noise seems to be a matter of speculation at the present time. Audiometric tests on one primitive tribe show that their hearing at seventy is comparable with that of Americans at thirty. In any event it becomes impossible to separate completely noise induced hearing loss from presbycousis in most cases. Typical presbycousis losses are shown in Fig. 2.11. These should apply to both men and women in the absence of noise induced hearing loss.

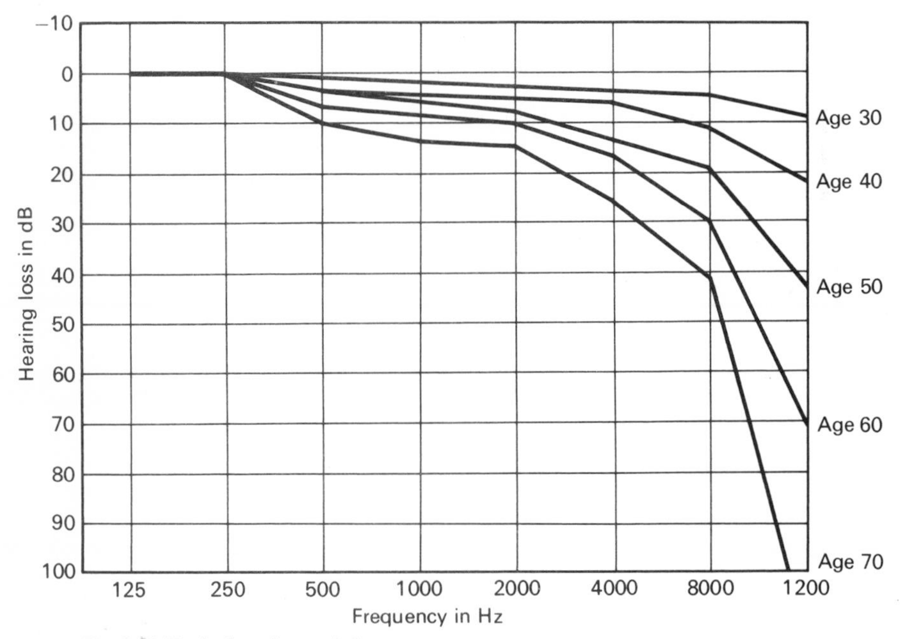

*Fig. 2.11* Typical presbycousis loss

*Tinnitus*
This defect which is experienced by most people from time to time is usually in the form of a high pitched ringing in the ears. Where people suffer frequently from tinnitus, they may even blame the noise on some other source, such as a piece of machinery.

*Deafness*
There are three main types of deafness: conductive deafness, nerve deafness, and cortical deafness.
*Conductive Deafness* is due to defects in those parts of the ear (external canal, ear drum and ossicles) which conduct the sound waves in the air to the inner ear. Examples are a thickening of the ear drum, stiffening of the joints of the ossicles or a blocking of the external canal by wax. These affect all frequencies evenly but the loss is limited to between 50 and 55 dB, due to conduction through the head.

Otosclerosis makes the stapes immobile. It can be helped by 'fenestration', where a new window is introduced into the lateral semi-circular canal.

A perforated ear drum can be caused by disease or by an explosion, but it can heal and recently artificial ear drums have been used.

Conductive hearing loss causes a loss despite amplitude. A 20 dB loss would mean that a sound of 20 dB is needed for threshold, 40 dB to hear a level of 20 dB, 60 dB to hear 40 dB etc. (at a particular frequency). People afflicted may not hear normal speech but may easily hear loud speech in a noisy factory.

*Nerve Deafness* is either due to loss of sensitivity in the sensory cells in the inner ear or to a defect in the auditory canal. There is no medical remedy and the hearing loss is usually different for different frequencies.

*Cortical Deafness* chiefly affects old people, and is due to a defect in the brain centres.

(2) Deafness Caused by Noise

It is not known exactly how loud a sound will cause immediate permanent deafness, but it is of the order of 150 dB. The harmful effect of long duration noise is even more difficult to assess. The Wilson report suggests the following maximum levels for more than 5 hours working per day in order to prevent any noticeable hearing loss.

**Table 2·3**

*Maximum Levels of Noise*

| Frequency Hz | 37·5–150 | 150–300 | 300–600 | 600–1200 | 1200–2400 | 2400–4800 |
|---|---|---|---|---|---|---|
| Value dB | 100 | 90 | 85 | 85 | 80 | 80 |

As it can be seen noise induced hearing loss depends on frequency. Narrow band noise is far more serious than broad band noise. A rough guide of 80 dB for the limit of 8 hours/day at 5 days/week for 40 years seems to be a reasonable level for normal noise.

*Temporary Threshold Shift and Permanent Threshold Shift*

When a person with normal hearing is exposed to intense noise for a few hours, he first suffers a temporary loss of hearing sensitivity called temporary threshold shift. After a sufficiently long rest from noise he usually recovers. In the case of a worker exposed to intense occupational noise during the working day for a matter of years, the stage is often reached where the temporary threshold shift has not completely recovered overnight before the next exposure. It appears that 'persistent threshold shift' is followed eventually by 'permanent threshold shift'. Research has shown that, in a given occupational noise the permanent threshold shift at 4000 Hz tends to become static after 10 years, when correction is made for ageing effects (presbycousis). Any further loss at 4000 Hz is due to ageing alone. Hearing at lower frequencies continues to deteriorate for longer periods.

In occupational surveys, hearing tests are conducted immediately after the annual leave, after a weekend or other absence from noise exposure to minimise the temporary threshold shift.

*Noise Susceptibility*

Most people have average noise susceptibility, but some appear to have 'tough' ears which suffer less from permanent threshold shift. Similarly there are those with 'tender' ears. In order to protect the latter, pre-employment tests followed by routine retests at frequent intervals are needed.

*Hearing Conservation and Surveys*

A hearing conservation programme is in three parts:

1. Analysis of noise exposure.
2. Control of noise exposure. (This may include reduction of noise at source, reduction of transmission of noise, ear protection — by ear plugs, muffs or change of job.)
3. Routine periodic audiometry.

Work is in progress to obtain data for the possibility of making occupational deafness a prescribed disease for the purpose of compensation under the Industrial Injuries Act.

## Other Physiological Effects of Noise

*Infrasound*

Infrasound consists of pressure waves with a frequency below 20 Hz and therefore below the normal threshold of hearing. Recently it has been established that infrasonic frequencies can occur in buildings with long ventilation ducts. A frequency of 7 Hz is particularly unpleasant and can cause a throbbing in the head due, it is suggested by Gavreau, to the fact that this coincides with the medium frequency of 'alpha' waves of the brain. Frequencies below 1 Hz are produced by wind effects on buildings.

*Vibration*

Human reaction to vibration depends upon both amplitude and frequency. It appears that the effect of vibration on the people within a building will be far more serious than the effect on the building. Tests that have been carried out show that similar painful human reactions have been produced by vibration of about 0·075 mm amplitude at 20 Hz and about 0·015 mm at 50 Hz.

Very large amplitudes of mechanical vibration, as with exceptionally high noise levels, may affect other sensory receptors such as touch.

It would appear that in general the subjective effect of vibration in buildings will set the limit for continuous vibration rather than structural considerations.

## Psycho-acoustics

Apart from the physiological effects of noise there is the psychological noise problem. This is even more difficult to define, because of the tremendous variability between individuals and environment. Music may sound enjoyable at a party but be annoying when someone is trying to work or sleep and perhaps be indirectly harmful to health when ill. In a factory the noise a machine produces may be helpful to its operator but have a dangerous and annoying masking effect on someone else. We might happily work in levels of 110 dBA in a machine shop, although our hearing may ultimately be damaged, but be disturbed by a tap dripping.

Various attempts have been made to correlate objective measures of a noise with its subjective effect on some sort of 'annoyance' or similar scale. The results obtained are usually only applicable for that particular type of noise in that particular environment. One of the following four types of scale is used: nominal, ordinal, interval or ratio. A nominal scale simply finds out that noise A is different from noise B, and therefore such a scale is of limited use. An ordinal scale determines the relative magnitudes. Interval scales are much more useful in that they show the size of graduations on the scale, but do not determine where the scale begins. A ratio scale finds the intervals and also the zero.

In one subjective test involving aircraft noise for the Committee on the Problem of Noise, people were asked to rate certain noises on a scale — not noticeable, noticeable, intrusive, annoying, very annoying or unbearable. This is an ordinal scale and not an interval scale because the sizes of the gaps between each of the named points on the scale are not known and are very unlikely to be equal. In this test it was, however, assumed to be an equal interval scale and later this was justified. It is very difficult and often impossible for subjects to rate noise on a ratio scale outside a laboratory, and to obtain even an interval scale may be difficult.

Another method of finding the effect of noise on people can be by determining the number of complaints that have been made about the particular noise. It is doubtful whether this relates to the average population who probably seldom complain, but it may be a useful guide to show the amount of annoyance.

Indirect methods may also be useful. One such method used to find the effect of motor traffic noise on people was to inquire whether householders slept at the front or rear of their house. The fact that many had moved from the main bedroom at the front to a smaller room at the rear indicated that the noise of traffic was having an annoying effect.

In practice, it may be necessary to use a combination of these methods. The present standards of sound insulation for dwellings are the results of social surveys based on them.

## Perceived Noise Level (PNdB)

The perceived noise level of a particular noise is the sound pressure level of a band of noise from 910 to 1090 Hz that sounds as 'noisy' as the sound under comparison. The noisiness is given in noys.

*Calculation of PNdB*

1. Find the maximum sound pressure level in each octave band centred at 63, 125, 250, 500, 1000, 2000, 4000 and 8000 Hz.
2. From Fig. 2.12 use the frequency and sound pressure level in dB to find the noisiness from the contours, e.g.: 80 dB for an octave centred at 1000 Hz gives a noy value of 17.
3. Find the total noisiness $N$ from

$$N = N_{max} + 0{\cdot}3(\Sigma N - N_{max})$$

where $N_{max}$ = highest noy value and
$\Sigma N$ = sum of the noy values in all octave bands.

   It will be seen that the formula is exactly equivalent to that used in the calculation of loudness, except for the use of equal noisiness contours in the place of equal loudness contours.
4. Noys may be converted to PNdB using

$$N = 2^{\frac{(x-40)}{10}}$$

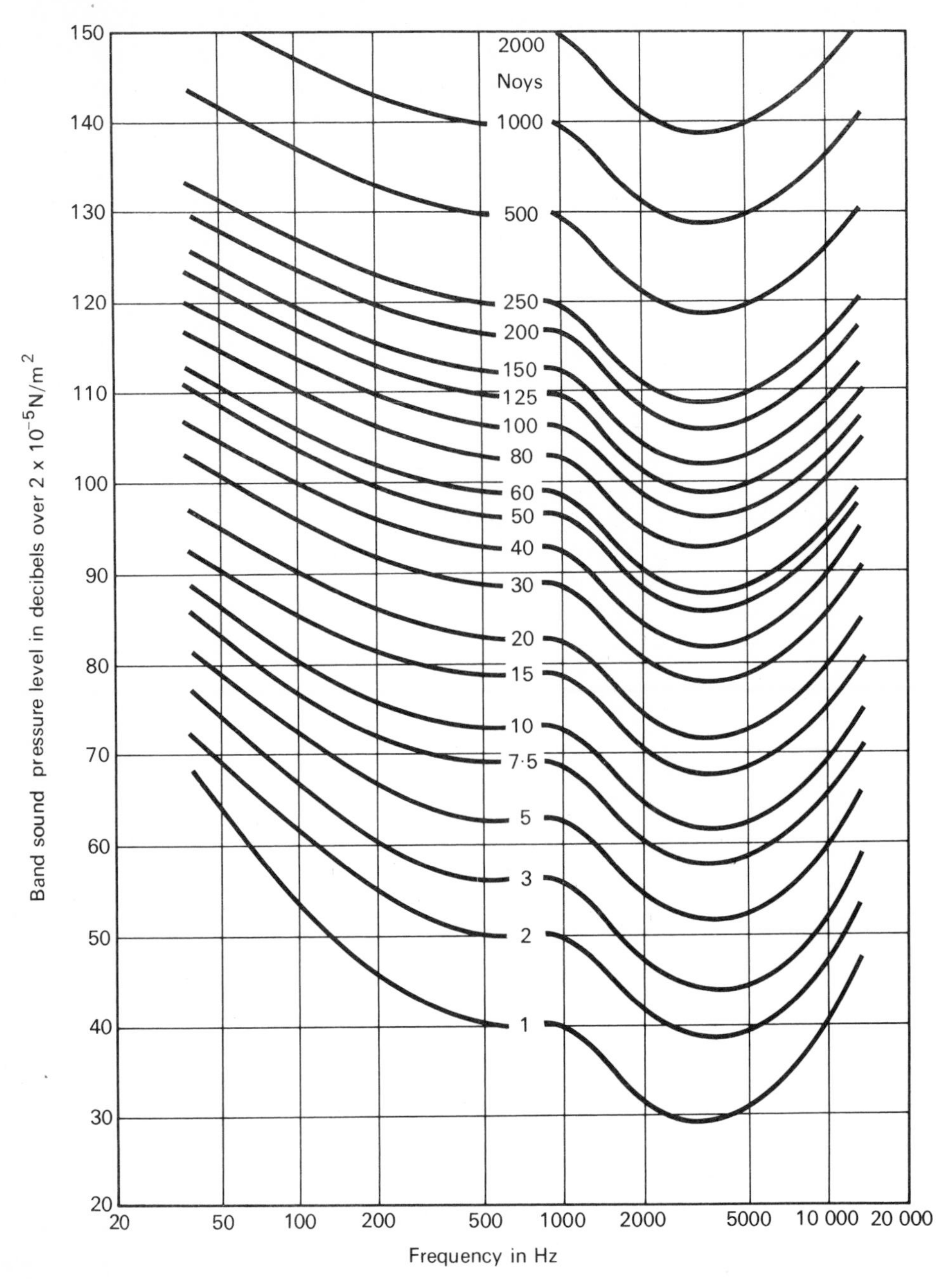


*Fig. 2.12* Contours of perceived noisiness

hence, $x = 40 + \frac{100}{3} \log_{10} N$

where $x$ = perceived noise level in PNdB.

Alternatively Table 2.4 may be used to find the value of $x$.
Note the similarity between sones for loudness and noys for noisiness, and between phons for loudness and PNdB for perceived noise level.

**Table 2·4**

*Perceived Noise Level as a Function of Total Perceived Noisiness From ISO (1966)*

| N | | | | N | | | |
|---|---|---|---|---|---|---|---|
| Lower | Mid | Upper | PNdB | Lower | Mid | Upper | PNdB |
| 1·0 | 1·0 | 1·0 | 40 | 15·5 | 16·0 | 16·6 | 80 |
| 1·1 | 1·1 | 1·1 | 41 | 16·7 | 17·1 | 17·7 | 81 |
| 1·1 | 1·1 | 1·2 | 42 | 17·8 | 18·4 | 19·0 | 82 |
| 1·2 | 1·2 | 1·3 | 43 | 19·1 | 19·7 | 20·4 | 83 |
| 1·3 | 1·3 | 1·4 | 44 | 20·5 | 21·1 | 21·8 | 84 |
| 1·4 | 1·4 | 1·5 | 45 | 21·9 | 22·6 | 23·4 | 85 |
| 1·5 | 1·5 | 1·6 | 46 | 23·5 | 24·2 | 25·1 | 86 |
| 1·6 | 1·6 | 1·7 | 47 | 25·2 | 26·0 | 26·9 | 87 |
| 1·7 | 1·7 | 1·8 | 48 | 27·0 | 27·8 | 28·8 | 88 |
| 1·9 | 1·9 | 1·9 | 49 | 28·9 | 29·8 | 30·9 | 89 |
| 2·0 | 2·0 | 2·1 | 50 | 31·0 | 32·0 | 33·1 | 90 |
| 2·1 | 2·1 | 2·2 | 51 | 33·2 | 34·3 | 35·5 | 91 |
| 2·3 | 2·3 | 2·4 | 52 | 35·6 | 36·8 | 38·1 | 92 |
| 2·5 | 2·5 | 2·5 | 53 | 38·2 | 39·4 | 40·8 | 93 |
| 2·6 | 2·6 | 2·7 | 54 | 40·9 | 42·2 | 43·7 | 94 |
| 2·8 | 2·8 | 2·9 | 55 | 43·8 | 45·2 | 46·8 | 95 |
| 3·0 | 3·0 | 3·1 | 56 | 46·9 | 48·5 | 50·2 | 96 |
| 3·2 | 3·2 | 3·4 | 57 | 50·3 | 52·0 | 53·8 | 97 |
| 3·5 | 3·5 | 3·6 | 58 | 53·9 | 55·7 | 57·7 | 98 |
| 3·7 | 3·7 | 3·9 | 59 | 57·8 | 59·7 | 61·8 | 99 |
| 4·0 | 4·0 | 4·1 | 60 | 61·9 | 64·0 | 66·3 | 100 |
| 4·2 | 4·3 | 4·4 | 61 | 66·4 | 68·6 | 71·0 | 101 |
| 4·5 | 4·6 | 4·7 | 62 | 71·1 | 73·5 | 76·1 | 102 |
| 4·8 | 4·9 | 5·1 | 63 | 76·2 | 78·8 | 81·6 | 103 |
| 5·2 | 5·3 | 5·5 | 64 | 81·7 | 84·4 | 87·4 | 104 |
| 5·6 | 5·6 | 5·8 | 65 | 87·5 | 90·5 | 93·7 | 105 |
| 5·9 | 6·1 | 6·3 | 66 | 93·8 | 97·0 | 100·4 | 106 |
| 6·4 | 6·5 | 6·7 | 67 | 100·5 | 104·0 | 107·6 | 107 |
| 6·8 | 7·0 | 7·2 | 68 | 107·7 | 111·4 | 115·3 | 108 |
| 7·3 | 7·5 | 7·7 | 69 | 115·4 | 119·4 | 123·6 | 109 |
| 7·8 | 8·0 | 8·3 | 70 | 123·7 | 128·0 | 132·5 | 110 |
| 8·4 | 8·6 | 8·9 | 71 | 132·6 | 137·2 | 142·0 | 111 |
| 9·0 | 9·2 | 9·5 | 72 | 142·1 | 147·0 | 152·2 | 112 |
| 9·6 | 9·8 | 10·2 | 73 | 152·3 | 157·6 | 163·1 | 113 |
| 10·3 | 10·6 | 10·9 | 74 | 163·2 | 168·9 | 174·8 | 114 |
| 11·0 | 11·3 | 11·7 | 75 | 174·9 | 181·0 | 187·4 | 115 |
| 11·8 | 12·1 | 12·5 | 76 | 187·5 | 194·0 | 200·8 | 116 |
| 12·6 | 13·0 | 13·5 | 77 | 200·9 | 207·9 | 215·3 | 117 |
| 13·6 | 13·9 | 14·4 | 78 | 215·4 | 222·8 | 230·7 | 118 |
| 14·5 | 14·9 | 15·4 | 79 | 230·8 | 238·8 | 247·3 | 119 |

### Example 2·2

Calculate the perceived noise level in PNdB of the motor car whose analysis is given in Example 2.1.

| Octave band | dB level | Noys |
|---|---|---|
| 20–75 | 95 | 21·0 |
| 75–150 | 84 | 13·8 |
| 150–300 | 80 | 14·0 |
| 300–600 | 68 | 7·5 |
| 600–1200 | 65 | 6·0 |
| 1200–2400 | 61 | 6·1 |
| 2400–4800 | 60 | 9·3 |
| 4800–10 000 | 60 | 16·5 |
| | $\Sigma N =$ | 94·2 |

$$N_{max} = 21{\cdot}0$$
$$\therefore N = 21{\cdot}0 + 0{\cdot}3(94{\cdot}2 - 21)$$
$$= 21{\cdot}0 + 0{\cdot}3(73{\cdot}2)$$
$$= 21{\cdot}0 + 21{\cdot}96$$
$$= 42{\cdot}96 \text{ noys}$$

From Table 2·4, 42·96 noys = 94 PNdB

### Noise and Number Index (NNI)

The noise and number index was originally derived to show the effect of aircraft noise on people. It is a composite measure taking into account both average peak noise level in PNdB of aircraft as well as the number of aircraft involved.

$$\text{NNI} = \text{average peak noise level} + 15 \log_{10} N - 80$$

where $N$ is the number of aircraft heard and average peak noise level is a logarithmic average, hence:

$$\text{average peak noise level} = 10 \log_{10} \frac{1}{N} \sum_{1}^{N} 10^{L/10}$$

and $L$ is in PNdB.

The Committee on the Problem of Noise found in the survey around London Heathrow Airport, that there was very good correlation between annoyance scores and the noise and number index. Thus it is possible to predict the amount of annoyance to be expected from peak noise levels and numbers of aircraft (see Fig. 2.13). Similarly it is possible to predict the noise and number index at various places at future times (see Figs. 2.14 and 2.15).

It would appear that there is some disagreement at the present time on the accuracy of prediction of the noise and number index.

### Relation of Noise Levels to Particular Environments

So far the effects of noise on people in general terms have been considered. However, it must be realized that the effects are not only due to the type of noise and the people listening, but also to their environment. For instance, it would be expected that in

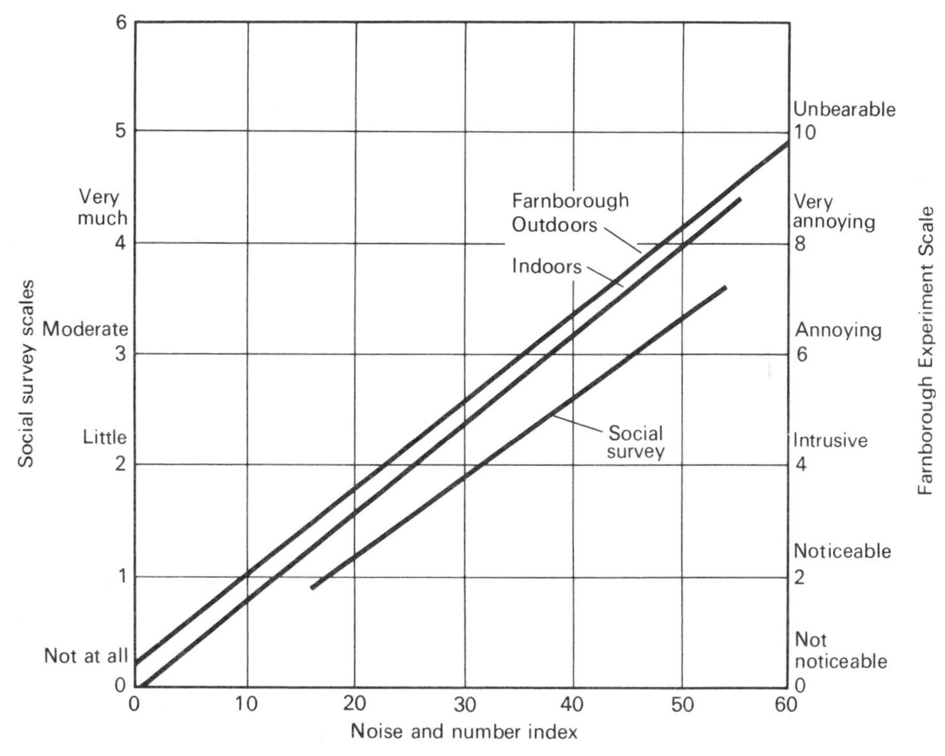


*Fig. 2.13* Relations between annoyance rating and noise and number index, obtained from social survey and Farnborough experiments

schools one of the most important factors is the effect that noises might have in masking speech. Conversely, in open plan offices or certain parts of hospitals this masking effect on speech is desirable.

The various criteria to be taken into account in different environments are discussed in the later chapters.

## Questions

(1) Calculate the loudness in phons of noise which has the following analysis:

| Octave band Hz | 20–75 | 75–150 | 150–300 | 300–600 | 600–1200 | 1200–2400 | 2400–4800 | 4800–10 000 |
|---|---|---|---|---|---|---|---|---|
| dB Level | 73 | 70 | 69 | 71 | 70 | 65 | 71 | 56 |

(2) If the noise in Question (1) is heard on the other side of a partition having the following sound reduction values, calculate the expected loudness in the receiving room.

| Octave band Hz | 20–75 | 75–150 | 150–300 | 300–600 | 600–1200 | 1200–2400 | 2400–4800 | 4800–10 000 |
|---|---|---|---|---|---|---|---|---|
| Insulation dB | 30 | 34 | 36 | 41 | 51 | 58 | 62 | 66 |

(3) Calculate the perceived noise level in PNdB of the noise in Question (1).

(4) Calculate the perceived noise level in PNdB of the noise in the receiving room in Question (2).

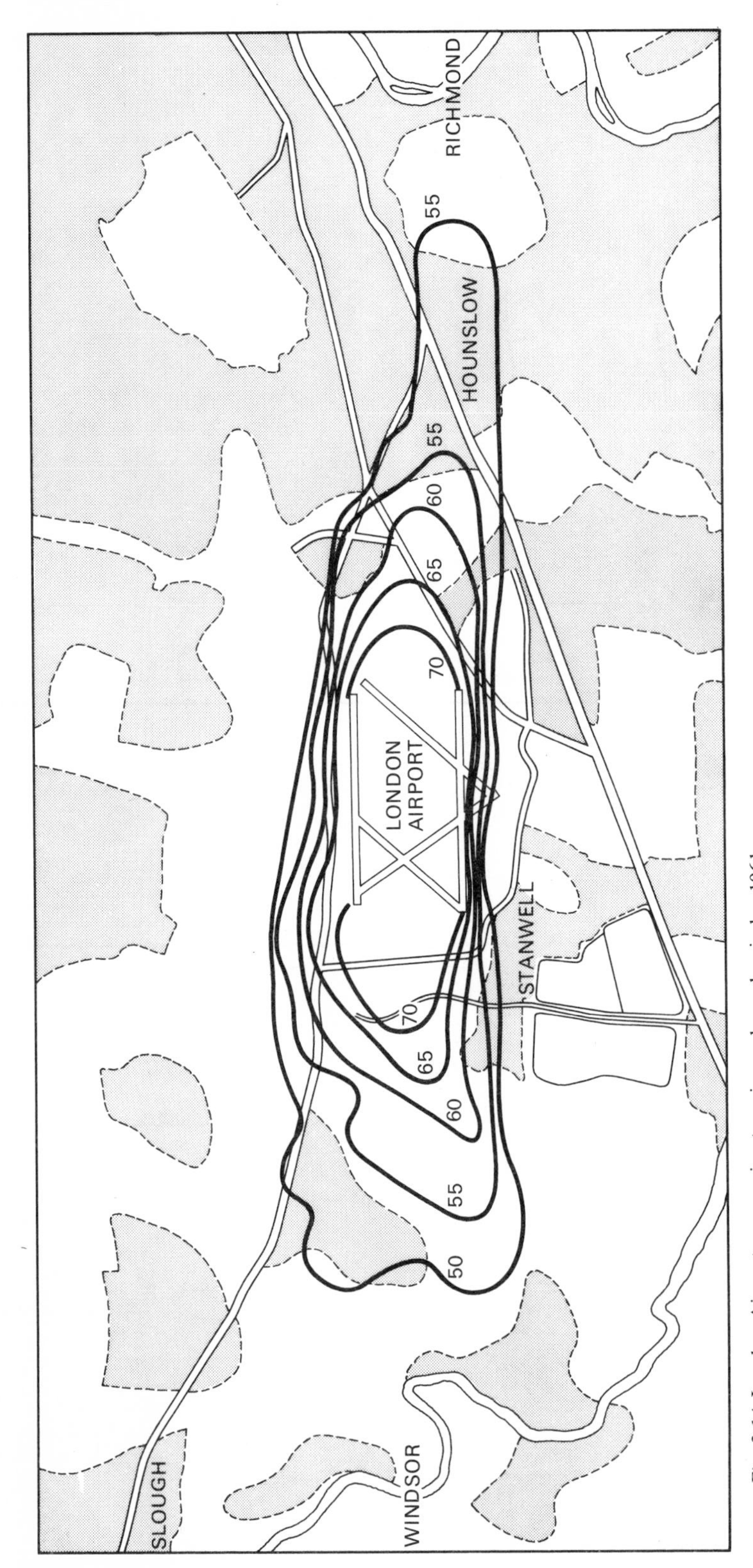


*Fig. 2.14* London Airport: approximate noise and number index 1961

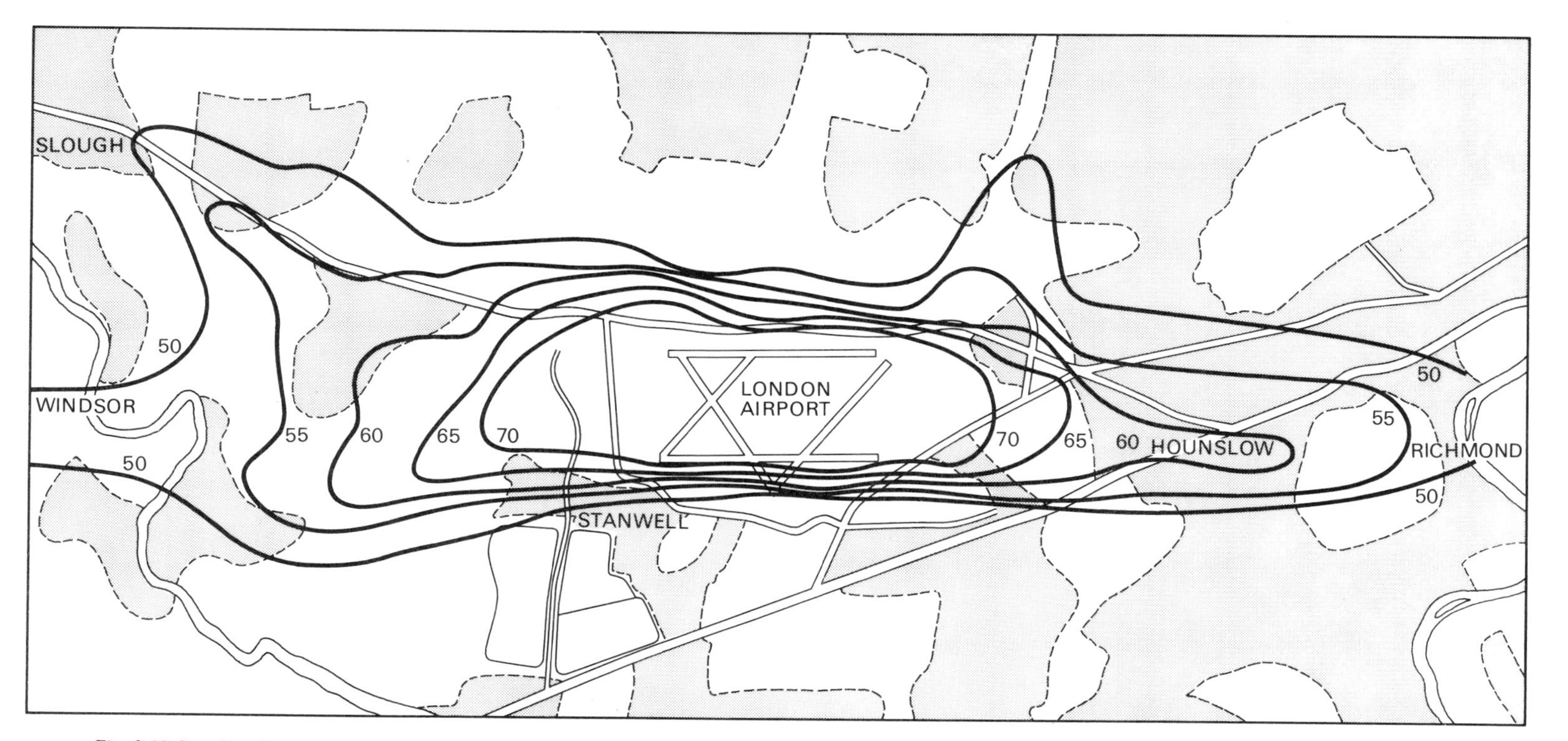


*Fig. 2.15* London Airport: estimated noise and number index 1970

(5) An octave band analysis of sound in a machine shop was made and the following results obtained:

| Octave band Hz | 20–75 | 75–150 | 150–300 | 300–600 | 600–1200 | 1200–2400 | 2400–4800 | 4800–10 000 |
|---|---|---|---|---|---|---|---|---|
| S.P.L. in dB | 68 | 72 | 90 | 87 | 86 | 88 | 90 | 84 |

Calculate the loudness in phons.

(6) Find the perceived noise level in PNdB of the analysis in Question (5).

# Room Acoustics

This chapter is concerned with the behaviour of sound within an enclosed space with a view to obtaining the optimum acoustic effect on the occupants. The various effects which rooms may have on the subjective properties of the sound are therefore studied.

## Requirements

1. An adequate amount of sound must reach all parts of the room. Most attention in this respect needs to be given to those seats furthest from the source.
2. An even distribution of sound throughout the room, irrespective of distance from the source.
3. Other noise which might tend to mask the required sound must be reduced to an acceptable level in all parts of the room.
4. The rate of decay of sound within the room (reverberation time) should be the optimum for the required use of the room. This is to ensure clarity for speech or 'fullness' for music.
5. Acoustical defects to be avoided include:
   (a) Long delayed echoes
   (b) Flutter echoes
   (c) Sound shadows
   (d) Distortion
   (e) Sound concentrations.

## Behaviour of Sound

*Reflection*

In the first chapter it was shown that sound can be reflected in a similar way to light, the angle of incidence being equal to the angle of reflection. However, it must be remembered that for this to be true, the reflecting object must be at least the same size as the wavelength concerned. It can often be very useful to carry out a limited geometrical analysis. This can prevent the problem of long delayed reflections, and focusing effects. It is impractical to take a geometrical analysis beyond the first or second reflections but it can prevent gross errors in design.

These errors are often caused by the focusing effects of concave shapes which may produce places with very loud sounds or dead spots, as shown in Fig. 3.1. It is generally unwise to have concave surfaces in a hall unless the focus is well outside. Convex surfaces can be useful in providing a diffusing surface in order to reflect the sound evenly in the hall.

*Long Delayed Reflections*

In large halls care must be taken to make certain that no strong reflections of sound are received by the audience after about 50 ms, otherwise confusion is likely between

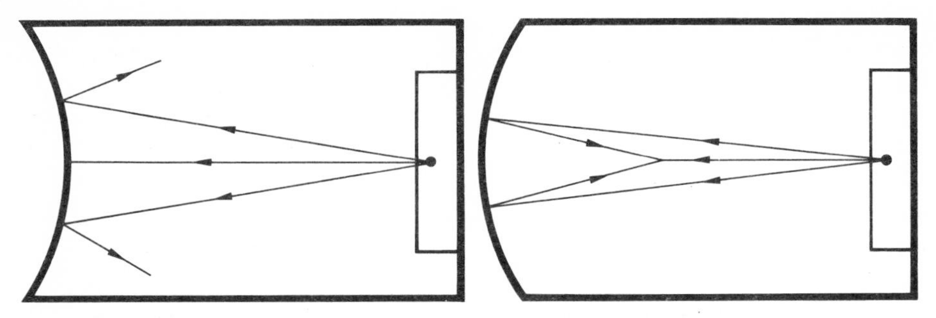

*Fig. 3.1* Diffusing effect of convex surfaces and focussing effect of concave surfaces

the direct and the reflected sound for speech. Average speech is at the rate of about 15 to 20 syllables per second or roughly 1 syllable every 70 ms to 50 ms respectively. This corresponds to a delay of about 17 metres. A member of the audience sitting at 8·5 metres from a good reflecting rear wall of a hall will find it difficult to understand speech. In larger halls other surfaces become important, such as the side walls and ceiling, as shown in Fig. 3.2. These strong reflections can be prevented by covering the surfaces concerned with absorbent material or by making them into diffusing surfaces by means of a convex shape. Quick reflections from the corners can be a problem but are easily overcome by the use of an acoustic plaster or some other absorbent material.

The simple solution appears to be to cover as much of the surfaces as possible with an absorbent material. Too much, however, would lead to a very unpleasant effect and also make speech without amplification too quiet to hear near the back of the hall. The aim should be to use the minimum amount of absorbent material so that the hall can have the minimum volume for a given number of people.

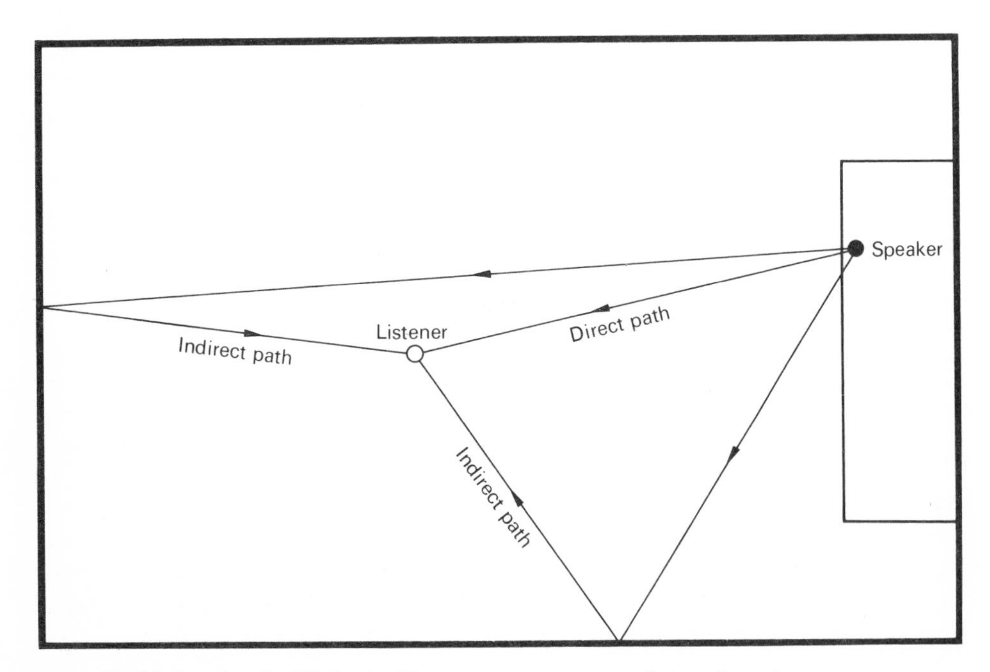


*Fig. 3.2* Sound paths differing by 17 m or more can cause confusion of speech

*Flutter Echoes*
These consist of a rapid succession of noticeable echoes which can be detected after short bursts of sound such as a hand clap (see Fig. 3.3). They can be avoided by making certain that the sound source is not between parallel reflecting surfaces. Sound absorbent material on one of the offending walls would also cure this defect, but the distribution of sound would become less uniform.

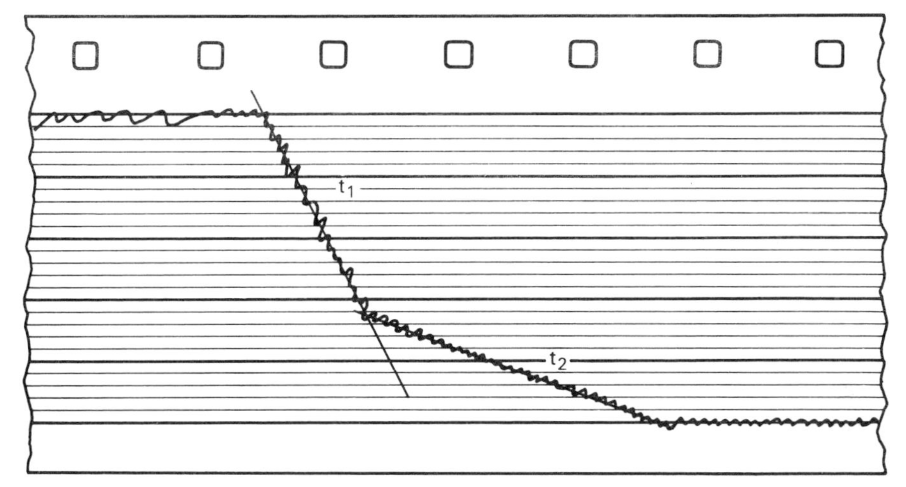


*Fig. 3.3* Recorded decay curve showing the effect of a flutter echo

Flutter echoes are particularly difficult to control in small rooms such as music practice rooms where much absorbent material is undesirable. The answer in this case is to avoid parallel walls and parallel ceilings and floors as shown in Fig. 3.4.

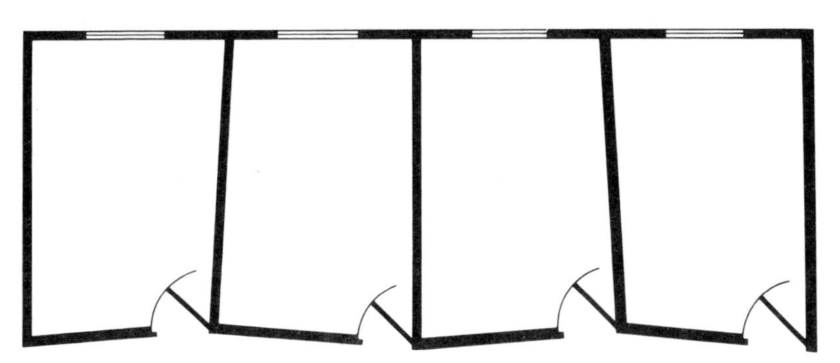

*Fig. 3.4* Flutter echoes can be avoided by making the walls a few mm out of parallel

## Ripple Tank Studies

The effect of awkward shapes can be studied using a ripple tank. This consists of a thin transparent tray of water in which the shape sits. The source of sound is a vibrator causing surface ripples. Effects of focusing, diffusion or interference can be investigated in this simple manner. A stroboscope shows this more clearly. It must be appreciated that this method only shows the effect for two dimensions and does not simulate absorption.

### Room Modes

In any particular room the behaviour of a sound is unique and highly complex because of the absorption and reflection properties of all the shapes within it. The fact that sound energy is not evenly distributed can be demonstrated by having a single frequency note played and listening to it while walking around the room. Positions of higher and lower levels will be noticed. Of course these positions are not the same for different frequencies. These variations in level should not be too great at audience listening positions.

### Acoustic Reflectors

Without amplification systems there will be a limited amount of sound power available. For unaided speech this is about 10 to 50 microwatts. In a larger hall the available sound power must be distributed evenly. There should be little problem near the front but further back it may be necessary to have sound directed by means of a specially shaped reflector. In many cases the ceiling can be utilized provided that long delayed reflections are prevented (see Fig. 3.5).

### Shape of Hall

Three basic plans are in common use for large halls. These are rectangular, fan and horse-shoe shaped. In a hall which seats under 1000 people the shape is not so critical. As the size increases there becomes a preference for the fan shape, so that the audience is seated slightly closer to the sound source (see Fig. 3.6). Care must be taken that the rear of such a hall is not concave. Problems can arise from reflections from side walls which may have to be broken up with either large diffusing surfaces or by the use of absorbent material. An example of a fan shaped hall is the F. R. Mann Concert Hall, Tel Aviv.

Opera houses are often built in a horse-shoe shape. The concave surfaces are broken up by tiers of boxes around the walls. The audience provides the absorption of sound. This type of finish is excellent for opera where clarity of sound is more important than fullness of tone but would not be considered so good for orchestral music. Covent Garden Opera House is an example of a horse-shoe shape.

The traditional rectangular shape has many advantages in construction and as long as reflectors are used over the sound source the difficulty of obtaining sufficient

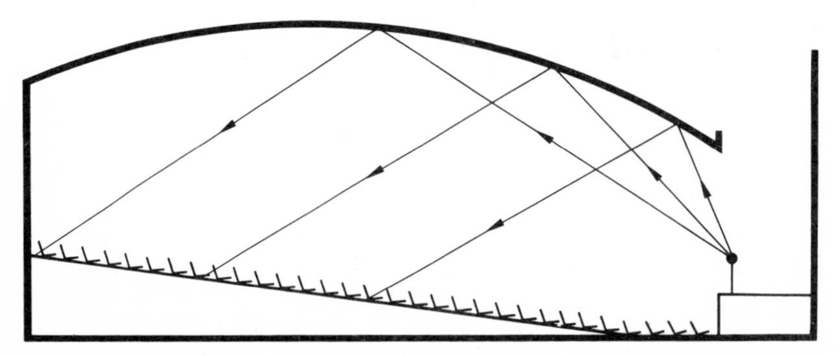

*Fig. 3.5* Correctly shaped ceiling or ceiling reflector can provide uniform distribution of sound energy

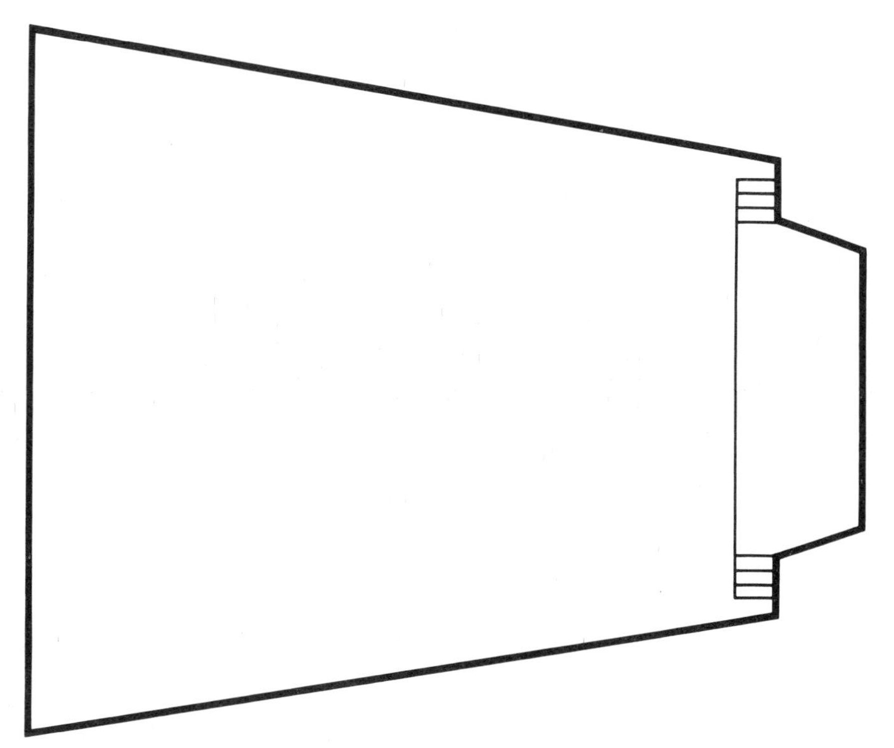

*Fig. 3.6* Fan-shaped hall

loudness near the back can be overcome. The Royal Festival Hall, London, is an example of a rectangular hall (see Fig. 3.7).

Other shapes have been used or suggested. Conventional shapes are probably more popular because the difficulties are less of an unknown quantity. The Royal Albert Hall is one example of an oval construction, but has acoustic problems. The Philharmonic Concert Hall, Berlin, is of irregular shape. The Philharmonic Hall, New

*Fig. 3.7* Rectangular-shaped hall

York, a combination of the rectangular and fan shape. Traditional dimensions have the ratio 2:3:5 for height:width:length.

Although traditionally rectangular in shape, churches have been built which are round. The problem of the focusing of sound by reflection has been overcome by careful use of absorbents.

### Seating Arrangements

Rows of people, particularly at grazing incidence to the sound, represent a most efficient absorber. It is essential in all but the smallest halls (above about 200 people) to rake the seating if at all possible. As a simple rule, adequate vision should ensure an adequate sound path. This will mean that the line of sight needs to be raised by 80 to 100 mm for each successive row (see Fig. 3.8).

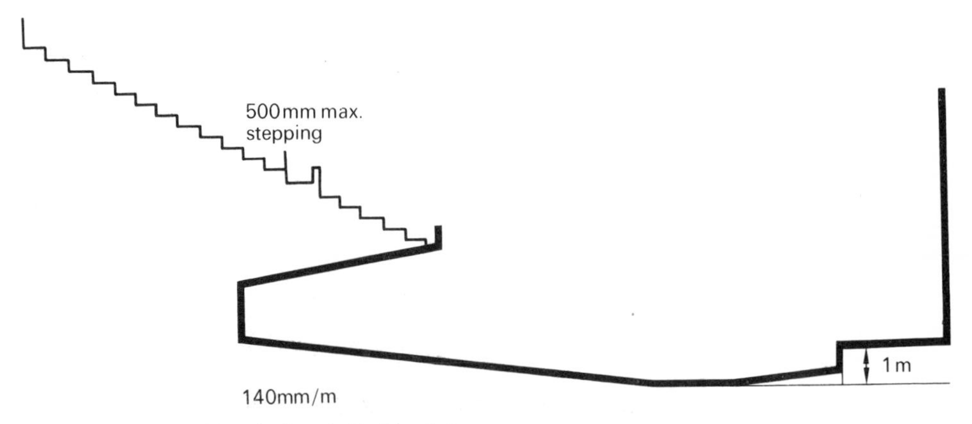


*Fig. 3.8* Suitable rake for a hall with a balcony

### Volume

In order that the optimum listening conditions are obtained, it is essential that a hall has the correct order of volume for its use. These are given in Table 3·1.

**Table 3·1**

*Optimum Volume/Person for Various Types of Hall*

| | Minimum | Optimum | Maximum |
|---|---|---|---|
| Concert Halls | 6·5 $m^3$ | 7·1 $m^3$ | 9·9 $m^3$ |
| Italian-type Opera Houses | 4·0 $m^3$ | 4·2–5·1 $m^3$ | 5·7 $m^3$ |
| Churches | 5·7 $m^3$ | 7·1–9·9 $m^3$ | 11·9 $m^3$ |
| Cinemas | — | 3·1 $m^3$ | 4·2 $m^3$ |
| Rooms for Speech | — | 2·8 $m^3$ | 4·9 $m^3$ |

It will be seen that the volume per person is dependent upon the purpose for which the building is to be used. Music played in a hall with too small a volume is likely to lack fullness, whereas speech in a hall with a very large volume for its seating capacity can be expected to lack clarity. Tables 3·2 and 3·3 give a list of the vital acoustic statistics for a few of the well known concert halls and opera houses. Nearly all of the best concert halls and opera houses in the world fit into the general pattern of volumes recommended in Table 3·1.

**Table 3·2**

*Acoustical Data for Some Well Known Concert Halls*

| Name | Volume $m^3$ | Audience capacity | Volume per aud. seat $m^3$ | Mid frequency R.T. in seconds (full hall) |
|---|---|---|---|---|
| St Andrew's Hall, Glasgow (built 1877) | 16 100 | 2133 | 7·6 | 1·9 |
| Carnegie Hall, New York (1891) | 24 250 | 2760 | 8·8 | 1·7 |
| Symphony Hall, Boston (1900) | 18 740 | 2631 | 7·1 | 1·8 |
| Tanglewood Music Shed Lennox, Mass. (1938) | 42 450 | 6000 | 7·1 | 2·05 |
| Royal Festival Hall (1951) | 22 000 | 3000 | 7·3 | 1·47 |
| Liederhalle, Grosser Saal, Stuttgart (1956) | 16 000 | 2000 | 8·0 | 1·62 |
| F. R. Mann Concert Hall, Tel Aviv (1957) | 21 200 | 2715 | 7·8 | 1·55 |
| Beethovenhalle, Bonn (1959) | 15 700 | 1407 | 11·2 | 1·7 |
| Philharmonic Hall, New York (1962) | 24 430 | 2644 | 9·3 | 2·0 |
| Philharmonic Hall, Berlin (1963) | 26 030 | 2200 | 11·8 | 2·0 |

**Table 3·3**

*Acoustical Data for Some Well Known Opera Houses*

| Name | Volume $m^3$ | Audience capacity | Volume per aud. seat $m^3$ | Mid frequency R.T. in seconds (full hall) |
|---|---|---|---|---|
| Teatro alla Scala, Milan (1778) | 11 245 | 2289 | 4·91 | 1·2 |
| Academy of Music, Philadelphia (1857) | 15 090 | 2836 | 5·32 | 1·35 |
| Royal Opera House (1858) | 12 240 | 2180 | 5·6 | 1·1 |
| Theatre National de L'Opera, Paris (1875) | 9960 | 2131 | 4·67 | 1·1 |
| Metropolitan Opera House, New York (1883) | 19 520 | 3639 | 5·36 | 1·2 |

Volumes of churches vary enormously, with many famous cathedrals having a capacity up to four times that recommended. St Paul's Cathedral, for example, has a volume of roughly 150 000 $m^3$. Many other English cathedrals have a volume of about 30 000 $m^3$ or more.

### Reverberation Time

This is the time it takes for a sound to decay by 60 dB.

One of the few concert hall criteria which can be defined precisely and measured with reasonable accuracy is the reverberation time. It is certainly one of the most important criteria. Sound does not die away the instant it is produced but will continue to be heard for some time because of reflections from walls, ceilings, floors and other surfaces. It will mix with later direct sound and is known as reverberant sound. People expect some reverberant sound which can assist understanding or help convey an atmosphere to an audience, from the haunted house with a long reverberation time to the padded cell. One's sense of well being or otherwise is undoubtedly affected by the length of the reverberation time.

Ideal reverberation times have been suggested by various workers using empirical methods. One such due to Stephens and Bate is:

$$t = r(0{\cdot}0118\sqrt[3]{V}+0{\cdot}1070)$$

where $t$ = reverberation time in seconds
$V$ = volume of the hall in $m^3$
$r$ = 4 for speech
= 5 for orchestra
= 6 for choir

An increase of about 40 per cent is advisable at the lower frequencies. Another method uses a set of graphs, as shown in Fig. 3.9. A few simple calculations show that these give similar results in most cases.

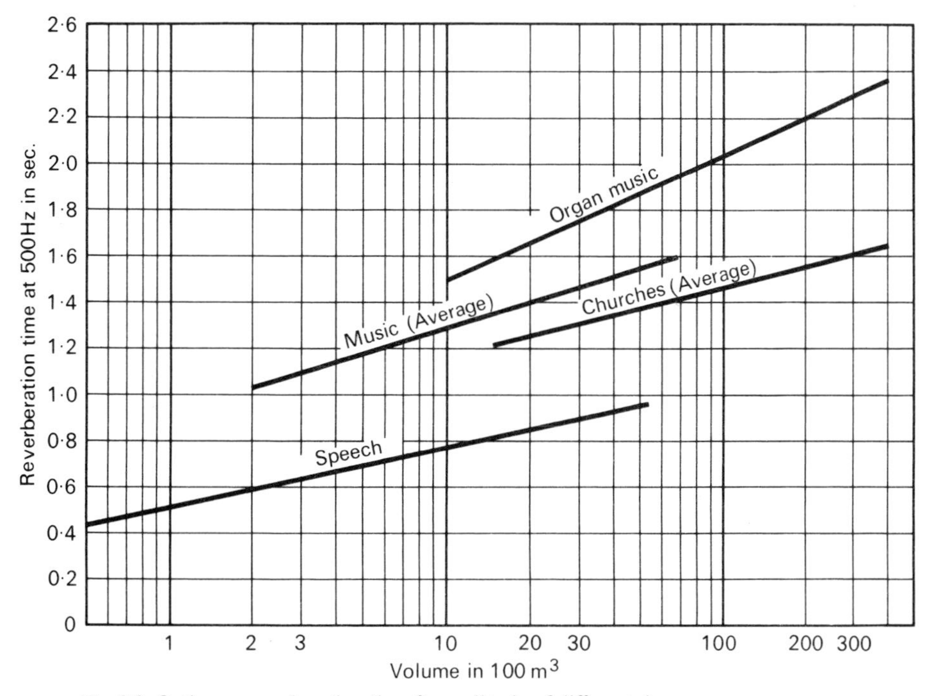


*Fig.* 3.9 Optimum reverberation time for auditoria of different sizes

### Example 3·1

Suggest the optimum volume and reverberation time for a concert hall to be used mainly for orchestral music and to hold 450 people.

From Table 3·1 a volume of 7·5 $m^3$ per person would be reasonable. This gives a total volume of 450 × 7·5 $m^3$ = 3375 $m^3$. Using the Stephens and Bate Formula:

$$t = r(0{\cdot}0118\sqrt[3]{V}+0{\cdot}1070)$$

where $r$ = 5 for orchestral music

$$\text{optimum reverberation time } t = 5(0{\cdot}0118\times\sqrt[3]{3375}+0{\cdot}1070)$$
$$= 5(0{\cdot}0118\times 15+0{\cdot}1070)$$
$$= 5(0{\cdot}2840)$$
$$= 1{\cdot}420\ s$$

*Sabine's Formula* — The actual reverberation time is calculated using the Sabine Formula:

$$t = \frac{0{\cdot}16\,V}{A}$$

where $t$ = reverberation time in seconds
$V$ = volume of hall in $m^3$
$A$ = absorption units in $m^2$

The number of absorption units is calculated using tables giving absorption coefficients for each of the materials used in the hall. The absorption coefficient for a material is the fraction of incident sound which is not reflected. Examples of these tables are given in Tables 3·4 and 3·5. The number of absorption units will normally be calculated for three or four selected frequencies — 125 Hz, 500 Hz, 2000 Hz, and if necessary 4000 Hz.

**Table 3·4**

*Absorption Coefficients of Common Building Materials*

| Material and method of fixing | Absorption coefficients | | | |
|---|---|---|---|---|
| | Low frequency 125 Hz | Medium frequency 500 Hz | High frequencies 2000 Hz | High frequencies 4000 Hz |
| Boarded roof; underside of pitched slate or tile roof | 0·15 | 0·1 | 0·1 | 0·1 |
| Boarding ('match') about 20 mm thick over air space on solid wall | 0·3 | 0·1 | 0·1 | 0·1 |
| Brickwork — plain or painted | 0·02 | 0·02 | 0·04 | 0·05 |
| Clinker ('breeze') concrete unplastered | 0·2 | 0·6 | 0·5 | 0·4 |
| Carpet (medium) on solid concrete floor | 0·1 | 0·3 | 0·5 | 0·6 |
| Carpet (medium) on joist or board and batten floor | 0·2 | 0·3 | 0·5 | 0·6 |
| Concrete, constructional or tooled stone or granolithic finish | 0·01 | 0·02 | 0·02 | 0·02 |
| Cork slabs, wood blocks, linoleum or rubber flooring on solid floor (or wall) | 0·05 | 0·05 | 0·1 | 0·1 |

| Material and method of fixing | Absorption coefficients | | | |
|---|---|---|---|---|
| | Low frequency 125 Hz | Medium frequency 500 Hz | High frequencies 2000 Hz | 4000 Hz |
| Curtains (medium fabrics) hung straight and close to wall | 0·05 | 0·25 | 0·3 | 0·4 |
| Curtains (medium fabrics) hung in folds or spaced away from wall | 0·1 | 0·4 | 0·5 | 0·6 |
| Felt, hair, 25 mm thick, covered by perforated membrane (viz. muslin) on solid backing | 0·1 | 0·7 | 0·8 | 0·8 |
| Fibreboard (normal soft) 13 mm thick mounted on solid backing | 0·05 | 0·15 | 0·3 | 0·3 |
| Ditto, painted | 0·05 | 0·1 | 0·15 | 0·15 |
| Fibreboard (normal soft) 13 mm thick mounted over air space on solid backing or on joists or studs | 0·3 | 0·3 | 0·3 | 0·3 |
| Ditto, painted | 0·3 | 0·15 | 0·1 | 0·1 |
| Floor tiles (hard) or 'composition' flooring | 0·03 | 0·03 | 0·05 | 0·05 |
| Glass; windows glazed with up to 3 mm glass | 0·2 | 0·1 | 0·05 | 0·02 |
| Glass; 7 mm plate or thicker in large sheets | 0·1 | 0·04 | 0·02 | 0·02 |
| Glass used as a wall finish (viz., 'Vitrolite') or glazed tiles or polished marble or polished stone fixed to wall | 0·01 | 0·01 | 0·01 | 0·01 |
| Glass wool or mineral wool 25 mm thick on solid backing | 0·2 | 0·7 | 0·9 | 0·8 |
| Glass wool or mineral wool 50 mm thick on solid backing | 0·3 | 0·8 | 0·75 | 0·9 |

**Table 3·4**—*Continued*
*Absorption Coefficients of Common Building Materials*

| Material and method of fixing | Absorption coefficients | | | |
|---|---|---|---|---|
| | Low frequency 125 Hz | Medium frequency 500 Hz | High frequencies | |
| | | | 2000 Hz | 4000 Hz |
| Glass wool or mineral wool 25 mm thick mounted over air space on solid backing | 0·4 | 0·8 | 0·9 | 0·8 |
| Plaster, lime or gypsum on solid backing | 0·02 | 0·02 | 0·04 | 0·04 |
| Plaster, lime or gypsum on lath, over air space on solid backing, or on joists or studs including decorative fibrous and plaster board | 0·3 | 0·1 | 0·04 | 0·04 |
| Plaster, lime gypsum or fibrous, normal suspended ceiling with large air space above | 0·2 | 0·1 | 0·04 | 0·04 |
| Plywood mounted solidly | 0·05 | 0·05 | 0·05 | 0·05 |
| Plywood panels mounted over air space on solid backing, or mounted on studs, without porous material in air space | 0·3 | 0·15 | 0·1 | 0·05 |
| Ditto, with porous material in air space | 0·4 | 0·15 | 0·1 | 0·05 |
| Water—as in swimming baths | 0·01 | 0·01 | 0·01 | 0·01 |
| Wood boards on joists or battens | 0·15 | 0·1 | 0·1 | 0·1 |
| Wood-wool slabs 25 mm thick (unplastered) solidly mounted | 0·1 | 0·4 | 0·6 | 0·6 |
| Wood-wool slabs 80 mm thick (unplastered) solidly mounted | 0·2 | 0·8 | 0·8 | 0·8 |
| Wood-wool slabs 25 mm thick (unplastered) mounted over air space on solid backing | 0·15 | 0·6 | 0·6 | 0·7 |

**Table 3·5**

*Absorption of Special Items*

| | Absorption units $m^2$ | | | |
|---|---|---|---|---|
| | Low frequency 125 Hz | Medium frequency 500 Hz | High frequencies 2000 Hz | High frequencies 4000 Hz |
| Air (per $m^3$) | — | — | 0·007 | 0·020 |
| Audience seated in fully upholstered seats (per person) | 0·19 | 0·47 | 0·51 | 0·47 |
| Audience seated in wooden or padded seats (per person) | 0·16 | 0·4 | 0·43 | 0·4 |
| Seats (unoccupied) fully upholstered (per seat) | 0·12 | 0·28 | 0·31 | 0·37 |
| Seats (unoccupied) wooden or padded or metal and canvas (per seat) | 0·07 | 0·15 | 0·18 | 0·19 |
| Theatre proscenium opening with average stage set (per $m^2$) | 0·2 | 0·3 | 0·4 | 0·5 |

**Example 3·2**

Calculate the reverberation time at 125 Hz, 500 Hz and 2000 Hz for a hall of volume 2500 $m^3$ having the following surface finishes:

| | |
|---|---|
| Plaster on brickwork | 265 $m^2$ |
| 3 mm glass window | 43 $m^2$ |
| Stage, boards on joist | 70 $m^2$ |
| 25 mm wood-wool slabs | 60 $m^2$ |
| Plate glass screen | 96 $m^2$ |
| Ceiling plaster | 310 $m^2$ |
| Wood block floor | 300 $m^2$ |

Assume the shading of the floor by the audience effectively reduces its absorption by 40 per cent at 125 Hz and 500 Hz, and by 60 per cent at 2000 Hz.

| Absorbent | | Absorption | | | | | |
|---|---|---|---|---|---|---|---|
| | | 125 Hz | | 500 Hz | | 2000 Hz | |
| Item | Area $m^2$ | Absorption coefficient | Absorption | Absorption coefficient | Absorption | Absorption coefficient | Absorption |
| Plaster | 265 | 0·02 | 5·3 | 0·02 | 5·3 | 0·04 | 10·6 |
| 3 mm glass | 43 | 0·3 | 12·9 | 0·1 | 4·3 | 0·05 | 2·2 |
| Stage, boards on joists | 70 | 0·15 | 10·5 | 0·1 | 7·0 | 0·1 | 7·0 |
| 25 mm wood-wool slabs | 60 | 0·1 | 6·0 | 0·4 | 24·0 | 0·6 | 36·0 |
| Plate glass | 96 | 0·1 | 9·6 | 0·04 | 3·8 | 0·02 | 1·9 |
| Ceiling plaster | 310 | 0·2 | 62·0 | 0·1 | 31·0 | 0·04 | 12·4 |
| Wood block floor minus shading | 300 | 0·05–40% | 9·0 | 0·05–40% | 9·0 | 0·1–60% | 12·0 |
| Audience | 250 | 0·17/ person | 42·5 | 0·43/ person | 107·5 | 0·47/ person | 117·5 |
| Air | 2500 $m^3$ | — | — | — | — | 0·01 | 25·0 |
| Total absorption | | | 157·8 | | 191·9 | | 224·6 |

$$\text{Total absorption at 125 Hz} = 157{\cdot}8\ \text{m}^2$$
$$\text{Total absorption at 500 Hz} = 191{\cdot}9\ \text{m}^2$$
$$\text{Total absorption at 2000 Hz} = 224{\cdot}6\ \text{m}^2$$

The actual reverberation time by Sabine's formula, $T = (0{\cdot}16\ V)/A$

$$T = \frac{0{\cdot}16 \times 2500}{157{\cdot}8} = 2{\cdot}5\ \text{s for 125 Hz}$$

$$\text{and } T = \frac{0{\cdot}16 \times 2500}{191{\cdot}9} = 2{\cdot}1\ \text{s for 500 Hz}$$

$$\text{and } T = \frac{0{\cdot}16 \times 2500}{224{\cdot}6} = 1{\cdot}8\ \text{s for 2000 Hz}$$

## Eyring's Formula

Sabine's formula has the great advantage of simplicity, and accuracy as long as the average absorption of all the surfaces within a room is less than about 0·2.

Sabine assumed that the sound within a room decayed continuously whereas Eyring considered intermittent decay at reflections.

Let $I_0$ be the original intensity of sound at time $t = 0$ and the average absorption coefficient of the reflecting surface be $\bar{\alpha}$.

$$\text{Therefore intensity after 1 reflection} = (1-\bar{\alpha})I_0$$
$$\text{intensity after 2 reflections} = (1-\bar{\alpha})^2 I_0$$
$$\text{intensity after } n \text{ reflections} = (1-\bar{\alpha})^n I_0$$

But reverberation time is the time taken for a 60 dB decay or a decay to $10^{-6}$ of the original intensity

$$\therefore \quad (1-\bar{\alpha})^n I_0 = 10^{-6} I_0$$
$$\therefore \quad (1-\bar{\alpha})^n = 10^{-6}$$
$$\therefore \quad n \log_e (1-\bar{\alpha}) = -6 \log_e 10$$
$$\text{or } n = \frac{-6 \log_e 10}{\log_e (1-\bar{\alpha})}$$

It can be shown that the mean free path in a rectangular room of volume $V$ and surface area $S = \dfrac{4V}{S}$ m (Reference 13). Hence the average time between reflections

$$= \frac{\frac{4V}{S}}{c}$$
$$= \frac{4V}{cS}$$

where $c$ = velocity of sound in air

$$\therefore \quad n \text{ reflections take a time } \frac{4Vn}{cS} \text{ s}$$

$$\therefore \quad \text{the reverberation time, } T = \frac{4V.(-6 \log_e 10)}{cS \log_e (1-\bar{\alpha})}$$
$$= \frac{-24 V \log_e 10}{cS \log_e (1-\bar{\alpha})}$$
$$= \frac{0{\cdot}16\, V}{-S \log_e (1-\bar{\alpha})} \qquad \left[ c = 330 \text{ m/s} \quad \frac{24 . \log_e 10}{330} = 0{\cdot}16 \right]$$

It should be noticed that the expansion of $\log_e (1-\bar{\alpha})$ is $-\bar{\alpha}-\bar{\alpha}^2/2-\bar{\alpha}^3/3-\bar{\alpha}^4/4$ etc. (Table 3·6) and for small values of $\bar{\alpha}$ (say 0·2) all terms after the first may be neglected. The reverberation time then becomes:

$$T = \frac{0{\cdot}16\, V}{(-S)(-\bar{\alpha})}$$
$$= \frac{0{\cdot}16\, V}{A}$$

A further correction may need to be added for higher frequencies to allow for air absorption.

$$\text{Then, } T = \frac{0{\cdot}16\, V}{-S \log_e (1-\bar{\alpha}) + xV}$$

where $x$ is the sound absorption/unit volume of air (Table 3·7). If the value of $\bar{\alpha}$ is less than about 0·2 but frequencies above 1000 Hz are being considered, then a modified form of Sabine's formula is convenient:

$$T = \frac{0{\cdot}16\,V}{A + xV}$$

**Table 3·6**

*Values of* $\log_e(1-\bar{\alpha})$ *corresponding to values of* $\bar{\alpha}$

| $\bar{\alpha}$ | $-\log_e(1-\bar{\alpha})$ | $\bar{\alpha}$ | $-\log_e(1-\bar{\alpha})$ | $\bar{\alpha}$ | $-\log_e(1-\bar{\alpha})$ |
|---|---|---|---|---|---|
| 0·01 | 0·0100 | 0·21 | 0·2355 | 0·41 | 0·5270 |
| 0·02 | 0·0202 | 0·22 | 0·2482 | 0·42 | 0·5441 |
| 0·03 | 0·0304 | 0·23 | 0·2611 | 0·43 | 0·5615 |
| 0·04 | 0·0408 | 0·24 | 0·2741 | 0·44 | 0·5792 |
| 0·05 | 0·0513 | 0·25 | 0·2874 | 0·45 | 0·5972 |
| 0·06 | 0·0618 | 0·26 | 0·3008 | 0·46 | 0·6155 |
| 0·07 | 0·0725 | 0·27 | 0·3144 | 0·47 | 0·6342 |
| 0·08 | 0·0833 | 0·28 | 0·3281 | 0·48 | 0·6532 |
| 0·09 | 0·0942 | 0·29 | 0·3421 | 0·49 | 0·6726 |
| 0·10 | 0·1052 | 0·30 | 0·3565 | 0·50 | 0·6924 |
| 0·11 | 0·1164 | 0·31 | 0·3706 | 0·51 | 0·7125 |
| 0·12 | 0·1277 | 0·32 | 0·3852 | 0·52 | 0·7331 |
| 0·13 | 0·1391 | 0·33 | 0·4000 | 0·53 | 0·7542 |
| 0·14 | 0·1506 | 0·34 | 0·4151 | 0·54 | 0·7757 |
| 0·15 | 0·1623 | 0·35 | 0·4303 | 0·55 | 0·7976 |
| 0·16 | 0·1742 | 0·36 | 0·4458 | 0·56 | 0·8201 |
| 0·17 | 0·1861 | 0·37 | 0·4615 | 0·57 | 0·8430 |
| 0·18 | 0·1982 | 0·38 | 0·4775 | 0·58 | 0·8665 |
| 0·19 | 0·2105 | 0·39 | 0·4937 | 0·59 | 0·8906 |
| 0·20 | 0·2229 | 0·40 | 0·5103 | 0·60 | 0·9153 |

**Table 3·7**

*x per m³ at a Temperature of 20°C*

| Frequency Hz | Relative Humidity, % 30 | 40 | 50 | 60 | 70 | 80 |
|---|---|---|---|---|---|---|
| | $\times 10^{-3}$ | $\times 10^{-3}$ | $\times 10^{-3}$ | $\times 10^{-3}$ | $\times 10^{-3}$ | $\times 10^{-3}$ |
| 1000 | 3·28 | 3·28 | 3·28 | 3·28 | 3·28 | 3·28 |
| 2000 | 11·48 | 8·2 | 8·2 | 6·56 | 6·56 | 6·56 |
| 4000 | 39·36 | 29·52 | 22·96 | 19·68 | 16·4 | 16·4 |

**Example 3·3**

Calculate the amount of absorption contributed at 2000 Hz by the 30 000 m³ of air in Coventry Cathedral when the relative humidity is 60 per cent. If its reverberation time empty at 2000 Hz is 4 s, find the number of m² of absorbent in the structure.

$$\begin{aligned} \text{Absorption} &= 6{\cdot}56 \times 10^{-3} \times 30\,000 \text{ m}^2 \\ &= 6{\cdot}56 \times 30 \\ &= 196{\cdot}8 \text{ m}^2 \end{aligned}$$

Using Sabine's formula:

$$T = \frac{0{\cdot}16\,V}{A + xV}$$

$$4 = \frac{0{\cdot}16 \times 30\,000}{A + 197}$$

$$A + 197 = \frac{0{\cdot}16 \times 30\,000}{4}$$

$$\begin{aligned} A &= 4 \times 300 - 197 \\ &= 1200 - 197 \\ &= 1003 \text{ m}^2 \end{aligned}$$

### Measurement of Reverberation Time

The chief problem is the accurate measurement of very small times which may be as little as 0·5 s. Basically the method is very simple. A sound is produced of suitable amplitude and then the rate of decay of its different frequencies is measured.

Ideally, the source of sound will be warble tones produced by a beat frequency oscillator, or narrow band random noise from a white noise generator, through an amplifier to a pair of loudspeakers, as shown in Fig. 3.10. This may be inconvenient and an alternative is to use a pistol as shown in Fig. 3.11.

The measurement can be done by the method shown in Fig. 3.10 or Fig. 3.11. This consists of a microphone connected to a frequency analyser connected in turn to a level recorder. The level recorder will need to have a logarithmic potentiometer in the circuit to convert the pressure measurement into dB.

In each case measurements will usually be made in octave bandwidths, whose centre frequencies are 125 Hz, 250 Hz, 500 Hz, 1000 Hz, 2000 Hz, 4000 Hz, or in third octave bandwidths, whose centre frequencies are 100 Hz, 125 Hz, 160 Hz, 200 Hz, 250 Hz, 315 Hz, 400 Hz, 500 Hz, 630 Hz, 800 Hz, 1000 Hz, 1250 Hz, 1600 Hz, 2000 Hz, 2500 Hz, 3150 Hz, 4000 Hz. This is done by switching the filters in the frequency analyser. A typical result for a certain room can be seen in Table 3·8.

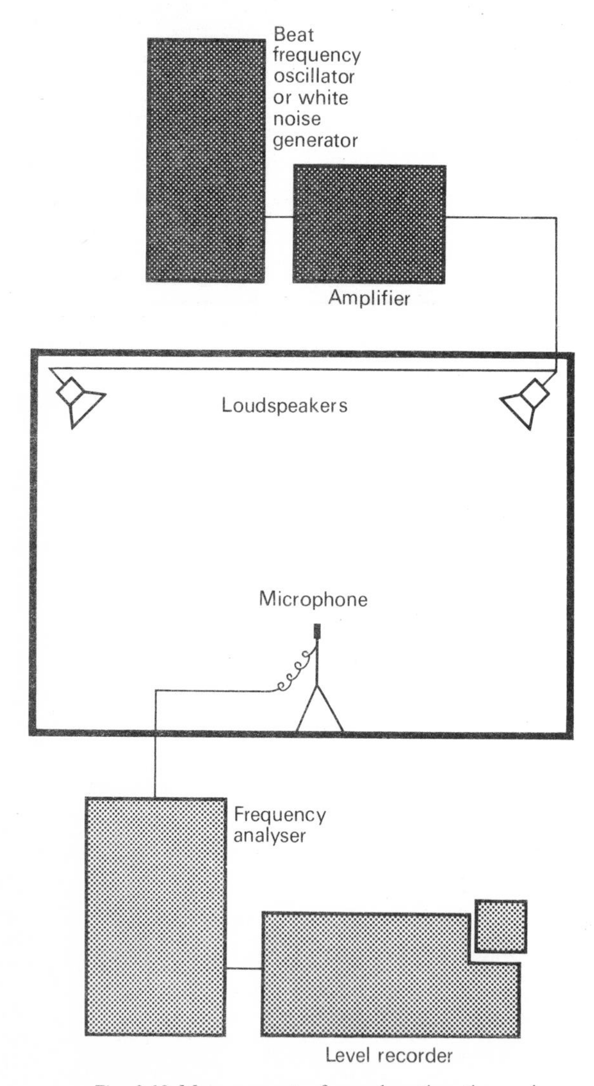


*Fig. 3.10* Measurement of reverberation time using a beat frequency oscillator or white noise generator as sound source

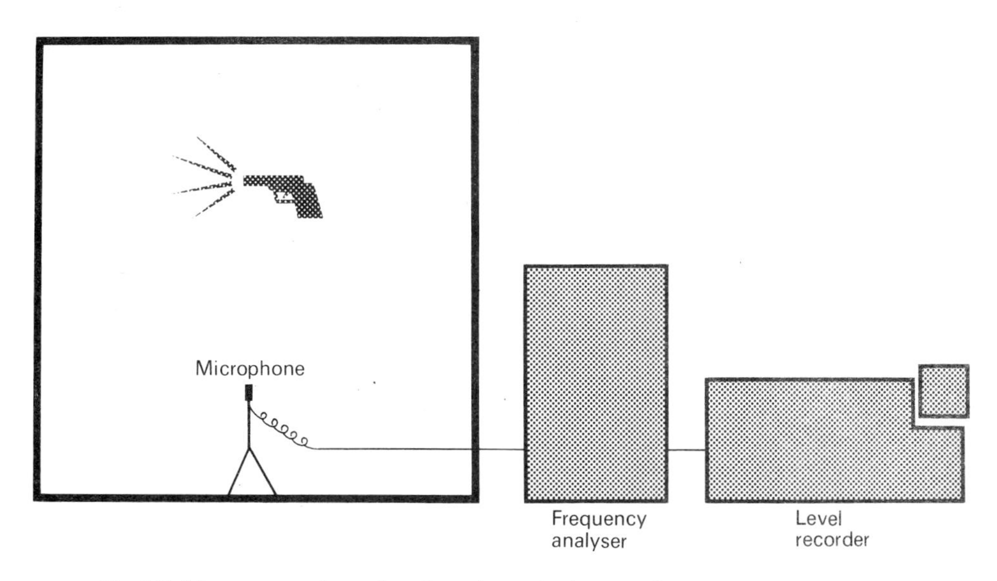


*Fig. 3.11* Measurement of reverberation using a pistol as sound source

**Table 3·8**

*Typical Reverberation Times for a Room*

| Third octave bandwidth centre frequency Hz | Reverberation time in s |
|---|---|
| 100 | 1·55 |
| 125 | 1·60 |
| 160 | 1·45 |
| 200 | 1·30 |
| 250 | 1·20 |
| 315 | 1·05 |
| 400 | 1·05 |
| 500 | 1·00 |
| 630 | 1·10 |
| 800 | 1·00 |
| 1000 | 0·90 |

**Table 3·8**—*Continued*
*Typical Reverberation Times for a Room*

| Third octave bandwidth centre frequency Hz | Reverberation time in s |
|---|---|
| 1250 | 1·05 |
| 1600 | 1·05 |
| 2000 | 1·05 |
| 2500 | 1·00 |
| 3150 | 0·95 |
| 4000 | 0·95 |

## Measurement of Absorption Coefficient

The Sabine formula shows that the reverberation time in a room is proportional to its volume and inversely proportional to absorption. It is essential that the absorption of the surface finishes in a hall are known at the design stage so that the expected reverberation time may be calculated. For this a value for the material known as absorption coefficient is needed. This is usually defined as the fraction of non-reflected sound energy to the incident sound energy. There are two main methods of measurement — the reverberation chamber or the impedance tube.

(1) Reverberation Chamber Method

This is the better method because it allows for all angles of incidence, but has the disadvantage of requiring a room of about 200 $m^3$ so that measurements down to 100 Hz may be made. (The lowest frequency should not be lower than about $125\sqrt[3]{(180/V)}$ Hz to ensure a diffuse sound field, where $V =$ the volume of the room.) The room itself will have walls and ceiling all slightly out of parallel. There will also be some diffusing objects in the form of curved sheets of plywood, or perspex a few mm thick. A long reverberation time is essential for accurate measurements.

The method is very simple. A measurement of reverberation time is made first without, and then with the absorbent material in the chamber.

$$\therefore \quad t_1 = \frac{0{\cdot}16\,V}{A}$$

and $$t_2 = \frac{0{\cdot}16\,V}{A+\delta A}$$

where $V$ = volume of reverberation chamber
$t_1$ = reverberation time of chamber
$t_2$ = reverberation time of chamber with absorbent material
$A$ = absorption of chamber without absorbent material
$\delta A$ = extra absorption due to the material

$$\therefore \quad A = \frac{0{\cdot}16\ V}{t_1} \tag{1}$$

$$A + \delta A = \frac{0{\cdot}16\ V}{t_2} \tag{2}$$

$$\therefore \quad \delta A = 0{\cdot}16\ V\left[\frac{1}{t_2} - \frac{1}{t_1}\right]$$

In practice some slight correction needs to be made for the behaviour of sound in the chamber which can make a difference of nearly 5 percent. These corrections are given in BS 3638 : 1963 *Method for the Measurement of Sound Absorption Coefficients (ISO) in a Reverberation Room.*

Where $\delta A = \left(55{\cdot}3\frac{V}{c}\right)\left[\frac{1}{t_2} - \frac{1}{t_1}\right]$

and $c =$ velocity of sound in air

the absorption coefficient, $\alpha = \frac{\delta A}{S}$

where $S =$ surface area under measurement which should be a single area between 10 and 12 $m^2$.

(2) Impedance Tube Method

This method only measures the absorption coefficient at normal incidence. It is a useful indication of the sort of absorbent properties which a material may have. Its main use is in theoretical work, research work or in quality control for the production of acoustic absorbent materials.

Pure tones produced by an oscillator are used to excite the loudspeaker, as shown in Fig. 3.12, producing standing waves in the tube. Partial reflection will take place at the absorbent surface resulting in a standing wave pattern as shown in Fig. 3.13.

If the displacement at any time of the incident wave is represented by:

$$d_1 = a \sin(wt - kx) \qquad \left[k = \frac{2\pi}{\lambda}\right]$$

and the displacement of the reflected wave by:

$$d_2 = fa \sin(wt + kx)$$

where $a =$ initial maximum amplitude

$fa =$ maximum amplitude of the reflected wave

Thus the resulting displacement at any point is given by:

$$\begin{aligned} d &= d_1 + d_2 \\ &= a \sin(wt - kx) + fa \sin(wt + kx) \\ &= a(1+f) \sin wt \cos kx + a(1-f) \cos wt \sin kx \end{aligned}$$

It can be seen that the maximum and minimum values will be $a(1+f)$ and $a(1-f)$ respectively, and $\lambda/4$ apart, the first being at 0, $\lambda/2$, $\lambda$, $3\lambda/2$, $2\lambda$, etc, the second at $\lambda/4$, $3\lambda/4$, $5\lambda/4$, $7\lambda/4$ etc.

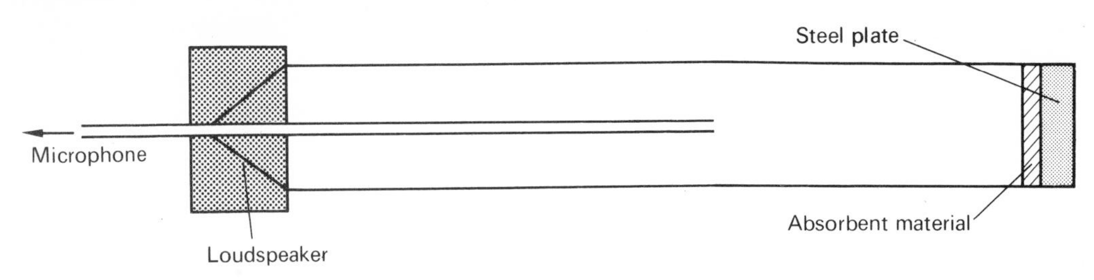


*Fig. 3.12* Impedance or standing wave tube for absorption co-efficient

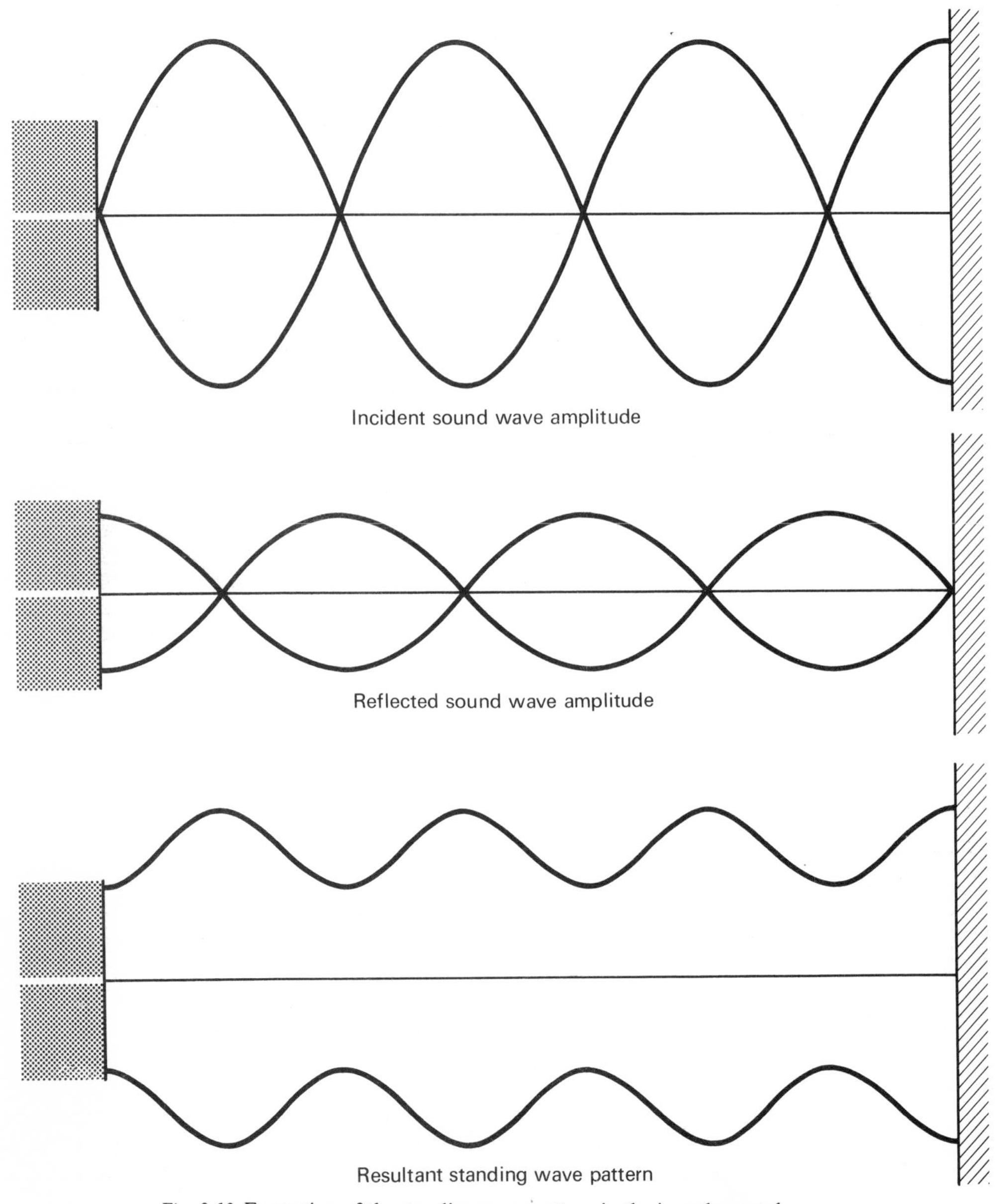


*Fig. 3.13* Formation of the standing wave pattern in the impedance tube

If the maximum and minimum amplitudes are $A_1$ and $A_2$

$$\text{then: } \frac{A_1}{A_2} = \frac{a(1+f)}{a(1-f)}$$

$$\text{or} \quad f = \frac{A_1 - A_2}{A_1 + A_2}$$

But the energy can be shown to be proportional to the square of the amplitude

$$\therefore \quad r = f^2 = \left[\frac{A_1 - A_2}{A_1 + A_2}\right]^2$$

where $r$ is the fraction of reflected energy.

The absorption coefficient, $\alpha = 1 - r$

$$\therefore \quad \alpha = 1 - \left[\frac{A_1 - A_2}{A_1 + A_2}\right]^2$$

$$= \frac{(A_1 + A_2)^2 - (A_1 - A_2)^2}{(A_1 + A_2)^2}$$

$$= \frac{2A_1 \,.\, 2A_2}{(A_1 + A_2)^2}$$

$$= \frac{4A_1A_2}{(A_1 + A_2)^2}$$

If the ratio of maximum :minimum, $[A_1/A_2]$ is measured, the formula is more conveniently written

$$\frac{4\left[\dfrac{A_1}{A_2}\right]}{\left[\dfrac{A_1}{A_2} + 1\right]^2} = \frac{4}{\left[\dfrac{A_1}{A_2} + \dfrac{A_2}{A_1} + 2\right]}$$

It will normally be found that results from the standing wave tube, though reproduceable, are less than those from the reverberation chamber. The size of the tube is also important. The maximum diameter of the sample should not be greater than about half of the wavelength under investigation. Thus, for measurements in a tube of 100 mm diameter, the upper limiting frequency is about 1600 Hz and for 6500 Hz the maximum diameter should be about 25 mm. It is possible to compare tube measurements of absorption coefficient with reverberation chamber measurements but the values in the latter can vary depending upon the distribution of the absorbent in the room.

### Types of Absorbents

Absorbents may be divided into three main types:

1. Porous materials
2. Membrane absorbers
3. Helmholtz resonators

The porous materials usually have some absorption at all frequencies, membrane absorbers have far more critical absorption characteristics around the resonant frequency of the panel, whilst Helmholtz resonator absorbers have even more critical absorption characteristics.

(1) Porous Absorbers

These consist of such materials as fibreboard, mineral wools, insulation blankets, etc. and all have one important thing in common — their network of interlocking pores. They act by converting sound energy into heat. Materials with closed cells such as the foamed plastics are far less effective as absorbers. Typical characteristics are shown in Fig. 3.14. Sound absorption is far more efficient at high than low frequencies. It may be slightly improved by increased thickness or mounting with an airspace behind. Porous absorbers are available in three types: prefabricated (tiles), plasters and spray on materials, and acoustic blankets (glasswool).

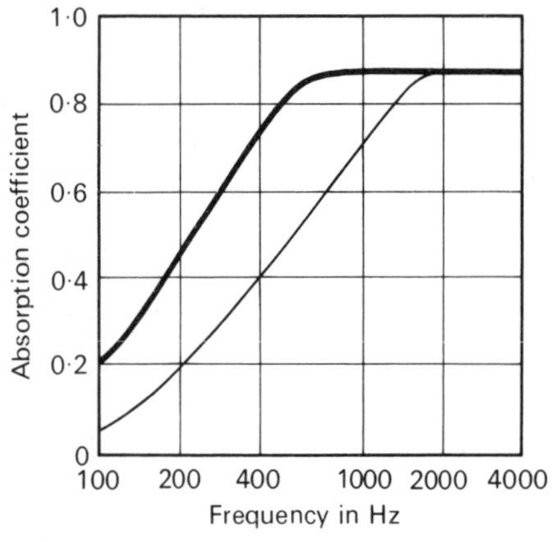


*Fig. 3.14* Typical absorption characteristics for porous materials showing (1) thin material, (2) thick material with its increase in absorption at lower frequencies

The method of fixing can make a considerable difference to the efficiency of these materials, particularly at low frequencies. In general it can be said that mounting the absorbent away from the wall surface results in a marked increase in low frequency absorption.

Where lack of space on walls or ceilings prevents the addition of absorbents they may be used in the form of space absorbers. These can be made from perforated sheets of steel, aluminum or hardboard in various shapes, such as cubes, prisms, spheres or cones and filled with glasswool or other suitable material. It is possible to make them with the underside reflecting while the top is absorbent. This can be very helpful in preventing long delayed sound from a dome in a hall reaching the listeners and at the same time providing more reflection of sound to certain parts of the audience.

(2) MEMBRANE OR PANEL ABSORBERS

These are useful because they can have good absorption characteristics in the low frequency range. The absorption is highly dependent upon frequency and is normally in the range 50 to 500 Hz (see Fig. 3.15).

The approximate resonant frequency, $f$, can be calculated from

$$f = \frac{60}{\sqrt{(md)}}$$

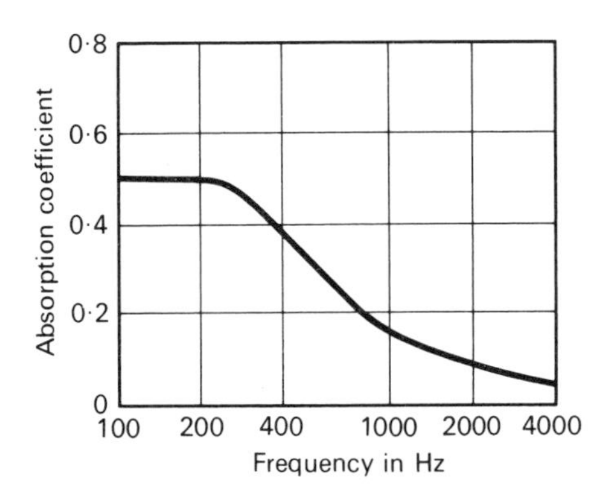


*Fig. 3.15* Typical absorption characteristics of a membrane absorber showing increased efficiency at low frequencies

where $m$ = mass of the panel in kg/m$^2$
and $d$ = depth of the air space in m

In practice this is only an approximation as the method of fixing and stiffness of individual panels can have a large effect. Panel absorbers can often be present fortuitously in the form of suspended ceilings or even closed double windows.

**Example 3·4**

A double glazed window with internal glass of mass 7 kg/m$^2$ has an air gap of 200 mm and is lined with an acoustic absorbent. Find the approximate expected resonant frequency.

$$f = \frac{60}{\sqrt{(7 \times 0{\cdot}2)}}$$

$$= \frac{60}{\sqrt{1{\cdot}4}}$$

$$= \frac{60}{1{\cdot}183}$$

$$= 50{\cdot}72 \text{ Hz}$$

(3) HELMHOLTZ OR CAVITY RESONATORS

These are containers with a small open neck and work by resonance of the air within the cavity. Porous material is introduced into the neck to increase the efficiency of absorption. It can be shown that for a narrow necked resonator of the type shown in Fig. 3.16 the resonant frequency, $f$, is approximately

$$f = \frac{cr}{2\pi}\sqrt{\left[\frac{2\pi}{(2l+\pi r)V}\right]}$$

where $c$ = velocity of sound in air
$r$ = radius of neck
$l$ = length of neck
$V$ = volume of cavity

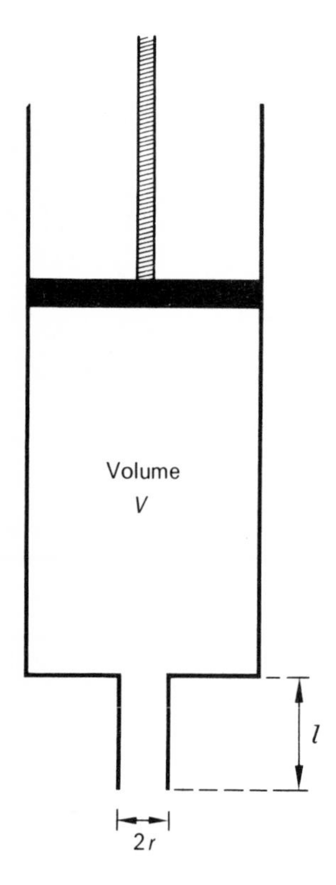


*Fig. 3.16* Helmholtz resonator

If there is no neck this reduces to:

$$f = \frac{c}{2\pi}\sqrt{\left(\frac{2r}{V}\right)}$$

Efficient absorption is only possible over a very narrow band, as shown in Fig. 3.17, and it is necessary to have many resonators tuned to slightly different frequencies if effective control of reverberation time is to be obtained. Cavity resonators are useful in controlling long reverberation times at isolated frequencies, and are used in a number of concert halls.

## Assisted Resonance

In theory, as has been shown, it is possible to design a hall with the ideal amount of absorption. In some cases variations in the amount of absorption with slight differences in construction can affect the final result. The optimum reverberation time is also dependent on the function of the hall (e.g. speech or music), so that some means of adjustment is useful.

One such system involves a time delay arrangement between a microphone and loudspeaker. The sound is played back at a suitably reduced amplitude after the necessary delay. The effect that must be conveyed subjectively is that of a steady decay of sound and not that of a delayed echo. Echo densities of 1000 per second or more may occur naturally in a hall. Subjectively these are interpreted as a continuous decay. Research done on artificial reverberation suggests that about 1000 echoes per second are needed for flutter-free reverberation and that the ear cannot distinguish any

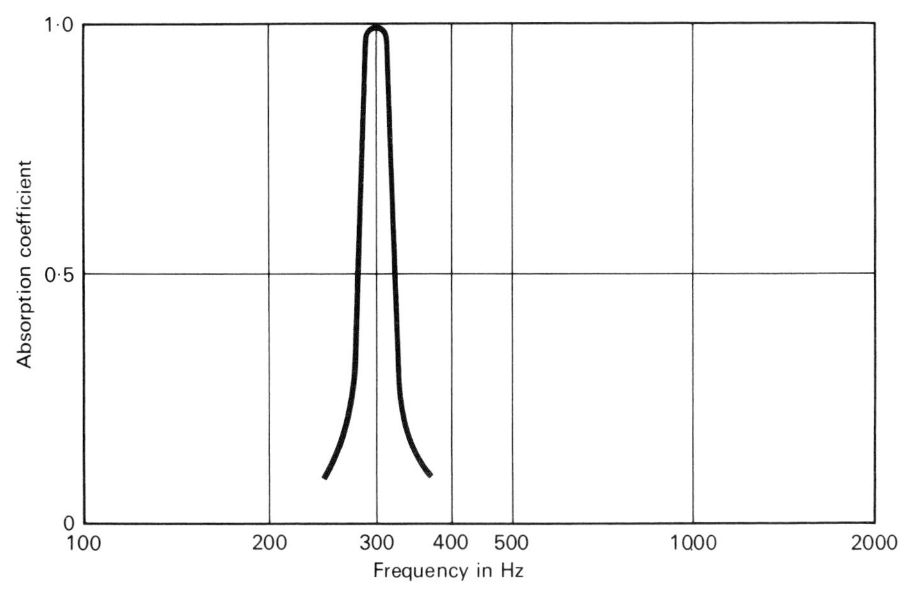


*Fig. 3.17* Characteristic very narrow band absorption by Helmholtz or cavity resonator

difference with a higher number. Echo densities of this magnitude are difficult to achieve economically. In addition it should be remembered that having an isolated loudspeaker from which reverberation is controlled would give the wrong spatial distribution. Thus it is necessary to have many loudspeakers each fed with different reverberated signals. This is known as ambiophonic reverberation. High quality equipment is needed throughout the whole chain of apparatus.

Another system which is in use in The Royal Festival Hall, appears to be very successful. This method uses a separate circuit for each very narrow band of sound. The circuit consists of a microphone within a Helmholtz resonator (Fig. 3.18) connected to an amplifier (and filter) which feeds a loudspeaker (Fig. 3.19). The microphone and loudspeaker are set up at similar maximum mode positions for their given frequencies and control is achieved by the gain of the amplifier. To avoid altering the natural acoustics of the hall other than the reverberation time it is important that the microphone and loudspeaker are exactly in phase or adjustment made to allow for this at the amplifier. In theory it should be possible to increase the natural reverberation time to any value up to infinity (the feed back position). Control of the system becomes difficult if the gain is near to the feed back position as shown in Fig. 3.20.

The Helmholtz resonator is acting as a narrow band filter with the result that high quality microphones and loudspeakers are not required as a linear response is not needed. For a hall with a reverberation time of approximately 1·5 s it appears that one channel is needed for about each 3 Hz at the lower frequencies decreasing to about one channel to 10 Hz at the higher frequencies.

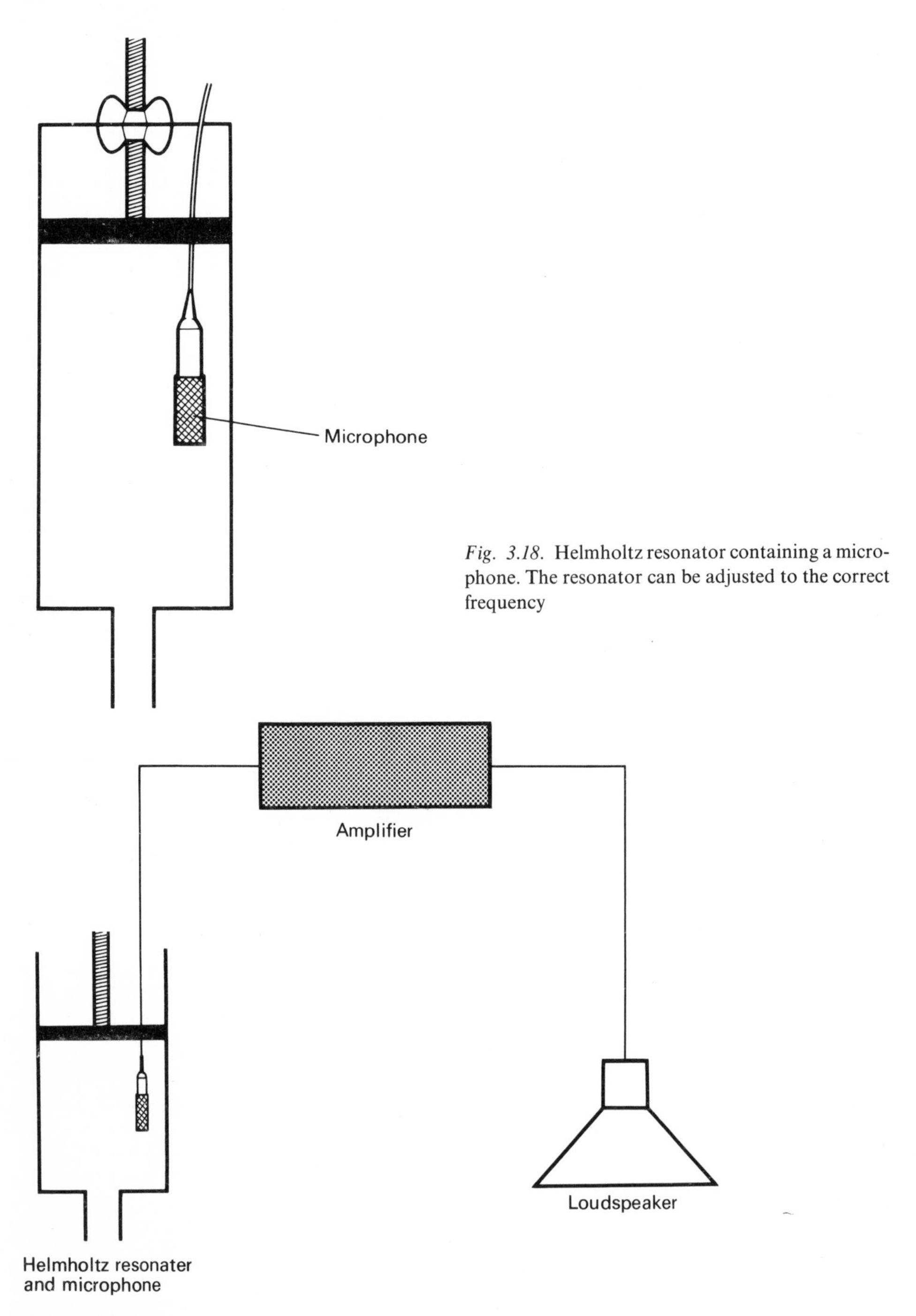


*Fig. 3.18.* Helmholtz resonator containing a microphone. The resonator can be adjusted to the correct frequency

*Fig. 3.19* Assisted resonance

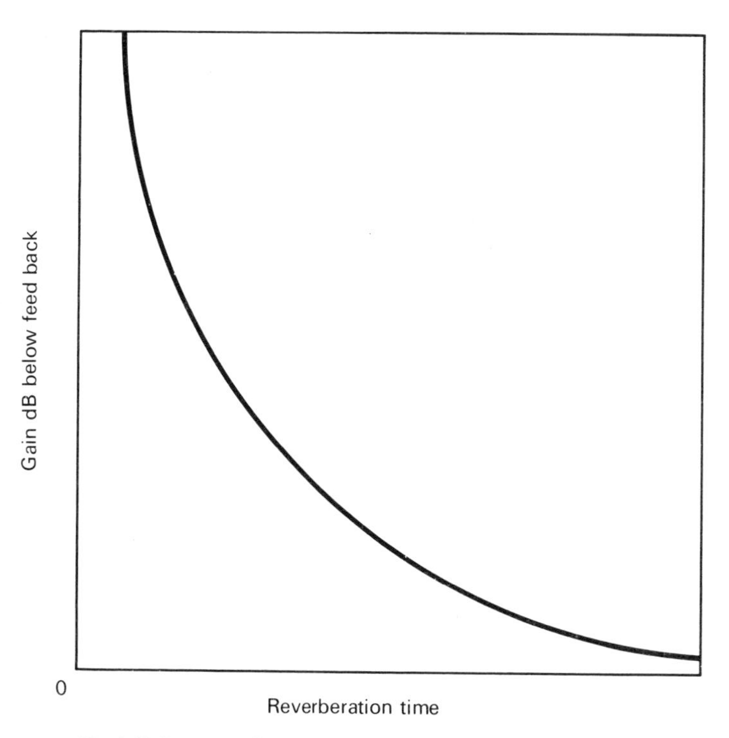


*Fig. 3.20* Increase of reverberation time at a given frequency for change in amplifier gain with assisted resonance

## Questions

(1) St Paul's Cathedral has a volume of about 150 000 $m^3$. Its reverberation time at 500 Hz is 11·7 s when empty, and 6·3 s when full. How many people would you expect to be present when it is full? (Assume that each person contributes 0·4 $m^2$ of absorption units.)

(2) A college lecture theatre is to be built to hold 200 people and will be used mainly for speech.
(a) Determine a suitable volume and reverberation time.
(b) How many absorption units would be needed in the construction to achieve optimum conditions when the hall is about two-thirds full? (Assume that each person contributes 0·4 $m^2$ of absorption units at 500 Hz.)

(3) A certain concert hall has a volume of 5·7 $m^3$ per person and holds 1800 people. The reverberation time is 1·6 s at mid frequencies. Assuming that Sabine's formula is accurate and that to achieve fullness of sound an orchestra requires one instrument for each 20 $m^2$ of absorption units, find the optimum size of orchestra.

(4) The absorption coefficient of a certain material was measured in a reverberation chamber of volume 1300 $m^3$ and the following average reverberation times were obtained:

| frequency Hz | R.T. in s (empty) | R.T. in s with 30 $m^2$ absorbent |
|---|---|---|
| 125 | 16·8 | 10·2 |
| 250 | 20·1 | 10·4 |
| 500 | 18·5 | 9·4 |
| 1000 | 14·5 | 8·0 |
| 2000 | 9·1 | 6·1 |

The average absorption coefficient of all the surfaces is less than 0·2. Find the absorption coefficients of the material at the frequencies given.

(5) Calculate the optimum and actual reverberation times at 500 Hz for a hall of volume 2500 m$^3$ so as to be satisfactory for choral music, using the following data.
Allow 40 per cent for shading of the floor.

*Data*:

| Item | Area (or No) m$^2$ | Absorption coefficient at 500 Hz |
|---|---|---|
| Wood block floor | 160 | 0·05 |
| Stage | 80 | 0·30 |
| Unoccupied seats | 50 | 0·15/seat |
| Audience | 200 | 0·4/person |
| Ceiling plaster | 160 | 0·1 |
| Canvas scenery | 96 | 0·3 |
| Perforated board | 120 | 0·35 |
| Glass | 40 | 0·10 |
| Plaster on brickwork | 200 | 0·02 |

How many additional absorption units are required to make the actual reverberation time equal to the optimum?

(6) (a) Calculate the optimum reverberation time for speech in a hall of volume 4000 m$^3$.
(b) Calculate the actual reverberation time at 500 Hz, in a hall with the following surface finishes and seating conditions:

| Item | Absorption coefficient at 500 Hz |
|---|---|
| 750 m$^2$ brick walls | 0·02 |
| 540 m$^2$ plaster on solid backing | 0·02 |
| 65 m$^2$ glass windows | 0·10 |
| 70 m$^2$ curtain | 0·40 |
| 130 m$^2$ acoustic board | 0·70 |
| 300 m$^2$ wood block floor | 0·05 |
| (allow 40 per cent for shading) | |

In addition there are 500 occupied seats each contributing 0·4 m$^2$ units of absorption. The volume of the hall is 4000 m$^3$.

(7) The reverberation time was measured in a lecture room of volume 150 m$^3$ and was found to be:

| Octave band centre frequency Hz | R.T. in s |
|---|---|
| 125 | 1·0 |
| 250 | 1·1 |
| 500 | 0·95 |
| 1000 | 1·0 |
| 2000 | 0·9 |
| 4000 | 0·8 |

Calculate the amount of extra absorption needed for each octave band.

(8) A room 16 m long, 10 m wide and 5 m high which was previously used as a laboratory is to be converted to use as a lecture room for 200 people. The original wall and floor surfaces are hard plaster and concrete whose average absorption coefficient is 0·05. Acoustical tiles of absorption coefficient 0·75 are available for wall or ceiling finishes. What is the desirable reverberation time for the new use of the room? Calculate the area of tile to be applied to achieve this. Absorption of seated audience (per person) is 0·4 m$^2$ units.

(9) (a) A hall 60 m long, 25 m wide and 8 m high has seating for 1200, and generally hard surfaces whose average absorption coefficient is 0·05. Calculate the reverberation time with a two thirds capacity audience for the frequency at which this data applies. The audience has an absorption of 0·4 $m^2$ units per person, but effectively reduces floor absorption by 40 per cent. The empty seats have an absorption of 0·28 $m^2$ units.
(b) What reduction in noise level would occur if the ceiling was then covered with acoustic tiles whose absorption coefficient is 0·6?

(10) A hall is to be built to hold about 600 people. Assuming that the normal use will be for speech, suggest the main points to be considered in order to obtain a suitable acoustic environment (without the use of amplification).

(11) A certain concert hall has a volume of 9 $m^3$ per person when full. It was designed to hold up to 1100 audience, 50 choir, and 35 orchestral players. If the ratio of volume/absorption is 12·5/1, how does the actual reverberation time full compare with the optimum suggested by the Stephens and Bate formula?

What would the reverberation time during a solo performance be in the absence of choir or orchestra if each person contributes 0·4 $m^2$ units of absorption?

(12) A lecture room 16 m long, 12·5 m wide and 5 m high has a reverberation time of 0·7 s. Calculate the average absorption coefficient of the surfaces using the Eyring formula.

Chapter 4

# Structure Borne Sound

Dilational waves in fluids (gases and liquids) within the audible frequency range can be detected by the human ear. This is obviously not the case with solids, although they may be detected by a sense of touch. Vibrations in solids do not therefore, in the strict sense, come within the category of sound. However, because of much mathematical similarity and also because these vibrations usually cause excitation of fluids in contact with the solids, it is convenient to refer to them as sound. In talking of structure borne or solid borne sound it is the effect of the solid coupling mechanism between source and hearer which is the chief consideration.

In this mechanism there are three important parts:

1. the coupling between the source of vibration and the solid structure.
2. the type of vibrational waves produced in the solid.
3. the coupling with the fluid at the hearing end.

A full mathematical treatment of the different types of wave propagation is not given in this chapter and readers interested in this are referred to the appropriate works on the subject. (Bibliographical references: 16, 17, 18, 19.)

## Wave Propagation in Solids

There are three main types of waves formed in solids, longitudinal, transverse and bending. In general the longitudinal waves are modified at all but high frequencies due to the limited thickness dimension of the solid. They do not cause much direct radiation of sound into the air, but are important because of their ability to excite other parts of the structure into bending vibrations. It is these bending waves which are the most important because of their ability to radiate sound.

(1) LONGITUDINAL WAVES

(a) *Velocity of Sound in a Bar*

Consider a thin uniform rod of cross-sectional area $A$ subject to longitudinal vibration (see Fig. 4.1). At some instant in time point $X$ is subject to a horizontal displacement $\eta$ from its original position $x$ from the end of the rod. In similar manner the displacement of point $Y$ is $\eta+\delta\eta$. The original distance between $X$ and $Y$ being $\delta x$.

$\therefore$ the strain at $X$ at this instant $= \dfrac{\partial \eta}{\partial x}$

and the rate of change of strain $= \dfrac{\partial^2 \eta}{\partial x^2}$

$\therefore$ the strain at $Y$ $= \dfrac{\partial \eta}{\partial x} + \dfrac{\partial^2 \eta}{\partial x^2} \cdot \delta x$

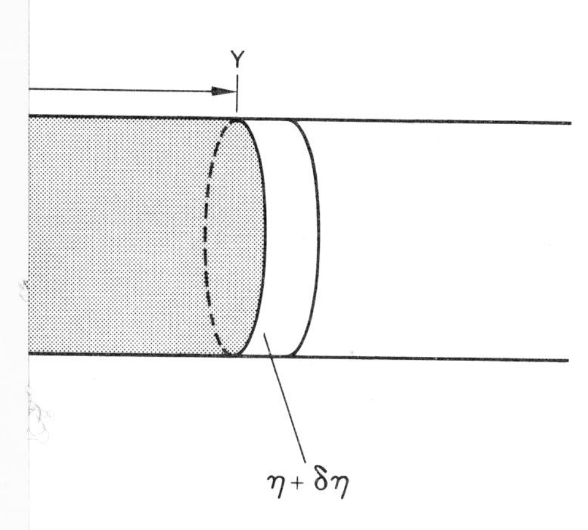


tions in a bar

$\therefore$ the stress at $X$ $= E\dfrac{\partial \eta}{\partial x}$

and the stress at $Y$ $= E\dfrac{\partial \eta}{\partial x} + E\dfrac{\partial^2 \eta}{\partial x^2} . \delta x$

$\therefore$ the change of stress $= E\dfrac{\partial^2 \eta}{\partial x^2} . \delta x$

$\therefore$ the force between the two points $= A . E . \dfrac{\partial^2 \eta}{\partial x^2} . \delta x$

By Newton's Law, force = mass × acceleration

$$= \rho . A . \delta x \frac{\partial^2 \eta}{\partial t^2}$$

where

$\rho$ = density
$t$ = time
$A$ = cross-sectional area
$E$ = Young's Modulus of Elasticity

$$\therefore \quad \rho . A . \delta x . \frac{\partial^2 \eta}{\partial t^2} = A . E . \frac{\partial^2 \eta}{\partial x^2} . \delta x$$

Hence it can be shown that

$$\rho \left[ \frac{\partial x}{\partial t} \right]^2 = E$$

But $\dfrac{\partial x}{\partial t} = V_L$, velocity of sound in the bar

$$\therefore \quad V_L^2 = \frac{E}{\rho}$$

or $$V_L = \sqrt{\left(\frac{E}{\rho}\right)}$$

In practice this relationship can be used to find Young's Modulus of Elasticity for materials such as concrete and glass. Table 4·1 gives values for the velocity of sound in some common materials.

**Table 4·1**

*Velocity of Sound in Some Common Materials*

| Material | Velocity in bar m/s |
|---|---|
| Glass | 5300 |
| Steel | 5200 |
| Aluminum | 5100 |
| Cast iron | 3400 |
| Reinforced concrete | 3700 |
| Bricks with mortar | 2350 |
| Wood — with grain<br>across grain | 3600–4900<br>1000–2500 |
| Rubber, hard<br>soft | 1400<br>50 |
| Sand | 97–210 |

(b) *Velocity of Sound in Plates*
The derivation is exactly similar to that in the preceding section, but taking into account the fact that the material will be restrained in another direction, and will thus be in a state of bi-axial stress. Under these conditions $E$ must be replaced by $E'$ where:

$$E' = \frac{E}{(1-\mu^2)}$$

where $\mu$ = Poisson's ratio

$$\therefore \quad V = \sqrt{\left[\frac{E}{\rho(1-\mu^2)}\right]}$$

(c) *Infinite Media*
In this case the material will be in tri-axial stress and the relationship becomes:

$$V = \sqrt{\left[\frac{E(1-\mu)}{(1+\mu)(1-2\mu)\rho}\right]}$$

It will be clear to the reader that this last case is not normally important in building structures. In the case of a brick wall of thickness 250 mm with a velocity in the region of 2500 m/s these conditions would only be approached at frequencies above 10 kHz.

The velocity of sound in solids will be of the order of ten times higher than in air, and Poisson's ratio will be of the order of 0·3 for most building materials. The latter makes little difference to the velocity.

(2) Transverse Waves

A solid is able to resist both change of volume and change in shape. As a result of the latter it can transmit tangential (shear) stresses. It can be shown that the velocity of propagation is:

$$V_T = \sqrt{\left[\frac{E}{2(1+\mu)\rho}\right]}$$

This gives a velocity of propagation very much lower than that for longitudinal waves. When $\mu = 0{\cdot}3$ the velocity of transverse waves will be about half that of longitudinal waves.

(3) Bending Waves

These are given a distinctive name because although transverse in appearance they are quite different from the waves that have just been considered (see Fig. 4.2). They are also by far the most important by virtue of their large deflections and hence their efficient radiating properties. There is a very small driving impedance. (Acoustic impedance is the ratio of sound pressure at the surface to the strength of the source.)

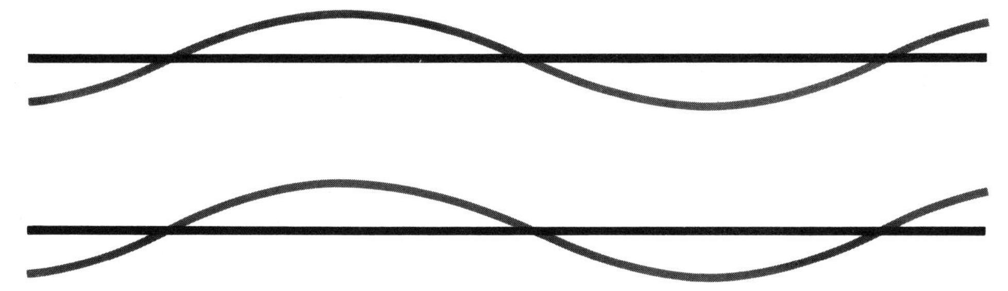

*Fig. 4.2* Bending waves produce large deflections in the structure and cause a considerable radiation of sound energy

The velocity of bending waves $V_B$ are frequency dependent and given by:

$$V_B = \left[\frac{EI}{A\rho}\right]^{1/4} . \omega^{1/2}$$

where $E$ = Young's modulus of elasticity
$I$ = second moment of area
$A$ = cross-sectional area of plate, beam or shell
$\rho$ = density
$\omega$ = frequency in radians/s

Alternatively this phase velocity of bending waves may be related to the longitudinal bar velocity, $V_L$ by:

$$V_B = (V_L\omega)^{1/2} . \left[\frac{I}{A}\right]^{1/4}$$

In the case of a flat plate this can be reduced to:

$$V_B = (1{\cdot}8hf\,V_L)^{1/2}$$

where $h$ = thickness of the plate

### Attenuation of Solid Borne Waves

Metals such as aluminium and steel have an attenuation less than air but greater than water. This means that for structural members there is very little inherent attenuation. Fortunately some is provided by joints and changes in dimension. Losses in building materials are often fairly large, particularly in certain granular and plastic materials.

Limited available data suggests that the attenuation by building materials such as concrete or brick is only of the order of $10^{-2}$ to $10^{-1}$ dB/m. In practice it would seem possible to obtain much larger losses from 1 to 3 dB/m due to other factors. These include friction at joints and changes of cross-section. Attenuation of bending waves is about half that of longitudinal waves.

### Changes of Cross-Section

The reduction factor is frequency independent for both longitudinal and bending waves. Figure 4.3 shows the reduction due to change in cross-section. It is fairly clear that for the orders of change that are practical in building construction the reduction is negligible.

### Corners

Reduction achieved by corners is dependent upon the fixing conditions. In the case of two identical beams joined by a swivel link at right-angles the reduction $R$, in dB can be shown to be:

$$R = 10\log_{10}\left[\frac{2\dfrac{V_B}{V_L}}{1+\dfrac{V_B}{V_L}+\dfrac{1}{2}\left[\dfrac{V_B}{V_L}\right]^2}\right]$$

$$= 10\log_{10}\left[\frac{2\left[1{\cdot}8\dfrac{hf}{V_L}\right]^{1/2}}{1+\left[1{\cdot}8\dfrac{hf}{V_L}\right]^{1/2}+0{\cdot}9\dfrac{hf}{V_L}}\right]$$

This reduction is reversible in that it applies to a longitudinal wave producing a bending wave or vice-versa.

If the two bars were rigidly joined then the mathematical analysis is far more complicated because there are six boundary conditions to be considered. When the two members are identical a 3 dB reduction results. Similarly at a T-junction a 6.5 dB reduction between the vertical and the arms of the T is found. In the case of X-junctions a 9 dB reduction is obtained.

It is fairly clear that the practical reduction achieved is quite small, and where compressional waves produce bending waves even of lower amplitude an increased level of room noise may result. A small reduction may be all that is needed. Thoughtful

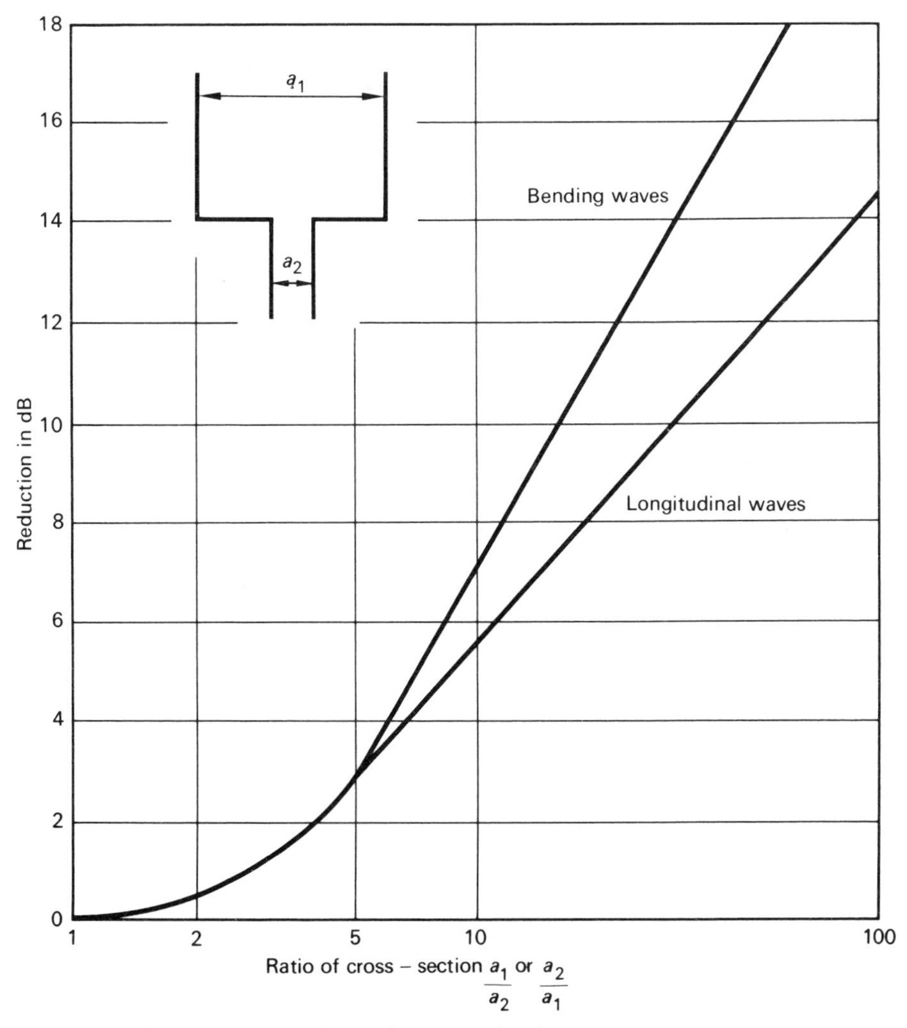


*Fig. 4.3* Reduction due to changes in cross-sectional area

positioning of domestic central heating pumps and convection preventing valves in relation to bends can prevent an annoying amount of noise being radiated into living rooms.

The reflection coefficient within a medium for normal incidence:

$$r = \frac{\text{reflected sound energy}}{\text{incident sound energy}}$$

$$= \frac{(R_2 - R_1)^2}{(R_2 + R_1)^2}$$

where $R_1$ and $R_2$ are the characteristic impedances of the two materials and may be calculated by the product $\rho V$ (density × velocity). Hence:

$$r = \left[\frac{\rho_2 V_2 - \rho_1 V_1}{\rho_2 V_2 + \rho_1 V_1}\right]^2 \quad \text{(see Table 4·2)}$$

**Table 4·2**

*Coupling of the Radiating Surface to Air*

| | Air 18°C | Water | Concrete | Iron | |
|---|---|---|---|---|---|
| Density | 1·2 | 1000 | 2300 | 7800 | $kg/m^3$ |
| Stiffness for longitudinal waves | $1{\cdot}4 \times 10^5$ | $2{\cdot}1 \times 10^9$ | $2{\cdot}3 \times 10^{10}$ | $2{\cdot}2 \times 10^{11}$ | $kg/m\ s^2$ |
| Longitudinal velocity | 342 | 1460 | 3160 | 5900 | m/s |
| Impedance | 410 | $1{\cdot}46 \times 10^6$ | $7{\cdot}3 \times 10^6$ | $4{\cdot}6 \times 10^7$ | $kg/m^2\ s$ |

It does not matter whether the sound is travelling from the more dense to a less dense medium or vice-versa. The applications are limited and the formula only applies under conditions where the two media may be considered infinite. The reason for this is that with thin sections, such as wall partitions, the reflected waves will have an interfering effect. If this is taken into account and the complex series summed then it can be shown that:

$$T = \frac{1}{1 + \left[\dfrac{\omega m}{2R_1}\right]^2}$$

where $T = 1 - r$, and is known as the transmission coefficient. It is normal to consider the transmission coefficient rather than the reflection coefficient.
$\omega$ = angular velocity in radians/s
$m$ = mass of the partition/unit area
$R_1 = \rho c$ = characteristic impedance of air

In practice this can be reduced to:

$$T = \frac{1}{\left[\dfrac{\omega m}{2R_1}\right]^2}$$ as the second term is large compared with 1.

The actual reduction is more conveniently put on a logarithmic scale in dB and becomes:

$$\text{Reduction in dB, } R = 10 \log_{10} \frac{1}{T}$$

$$= 10 \log_{10} \left[\frac{\omega m}{2R_1}\right]^2$$

$$= 20 \log_{10} \left[\frac{\omega m}{2\rho c}\right]$$

This shows that in theory doubling the frequency or the mass should have the same effect, about 6 dB better insulation. Unfortunately the theory does not agree with measurements made on many samples. In practice sound is incident at all angles and not just at right-angles.

For other angles of incidence to the normal, the transmission coefficient,

$$T_\theta = \frac{1}{1+\left[\dfrac{\omega m \cos\theta}{2\rho c}\right]^2}$$

By summing the transmission for $\theta$ from 0 to 90° it can be shown that the reduction of sound in dB for all angles of incidence,

$$R = R_0 - 10\log_{10} 0{\cdot}23\, R_0$$

where $R_0$ = reduction of sound at normal incidence

$$= 10\log_{10}\left[\frac{\omega m}{2\rho c}\right]^2$$

The result gives an increase of about 5 dB per doubling of frequency or mass per unit area. This is much nearer the measured value. In fact integration over the full range 0 to 90° is not quite correct as little sound would be present at grazing incidence.

## Resonance of Partitions

In the previous section it was assumed that a panel was completely controlled by its mass. This is certainly true for most practical purposes over much of its frequency range, which for many constructions is within limits of audibility. Below these frequencies a panel becomes largely stiffness controlled and acts in the manner of a spring. This means that there will be resonant frequencies which depend on stiffness and fixing conditions as well as mass.

These resonant conditions can be demonstrated by means of a rectangular metal plate clamped horizontally through a centre hole to a vibrator (Goodman or Advance). The frequency may be adjusted by a simple audio-frequency generator. Sand placed on the plate will form patterns at certain frequencies, the lowest resonant condition producing the simplest. If the edges are fixed the resonant frequencies are changed and a different pattern is formed (see Fig. 4.4).

If a rectangular panel is supported at its edges but not clamped, its resonant frequencies may be calculated from:

$$f = 0{\cdot}45 V_L h\left[\left(\frac{N_x}{l_x}\right)^2 + \left(\frac{N_y}{l_y}\right)^2\right] \text{Hz}$$

where $V_L$ = longitudinal velocity in m/s
$h$ = thickness in m
$l_x$ = length in m
$l_y$ = height in m
$N_x, N_y$ = integers 1, 2, 3, etc. The lowest, 1 and 1 giving the lowest resonant frequency

### Example 4·1

Calculate the lowest resonant frequency for a brick partition 120 mm thick, 4 m by 2 m in area with a longitudinal wave velocity of 2350 m/s. (Assume it is supported at its edges.)

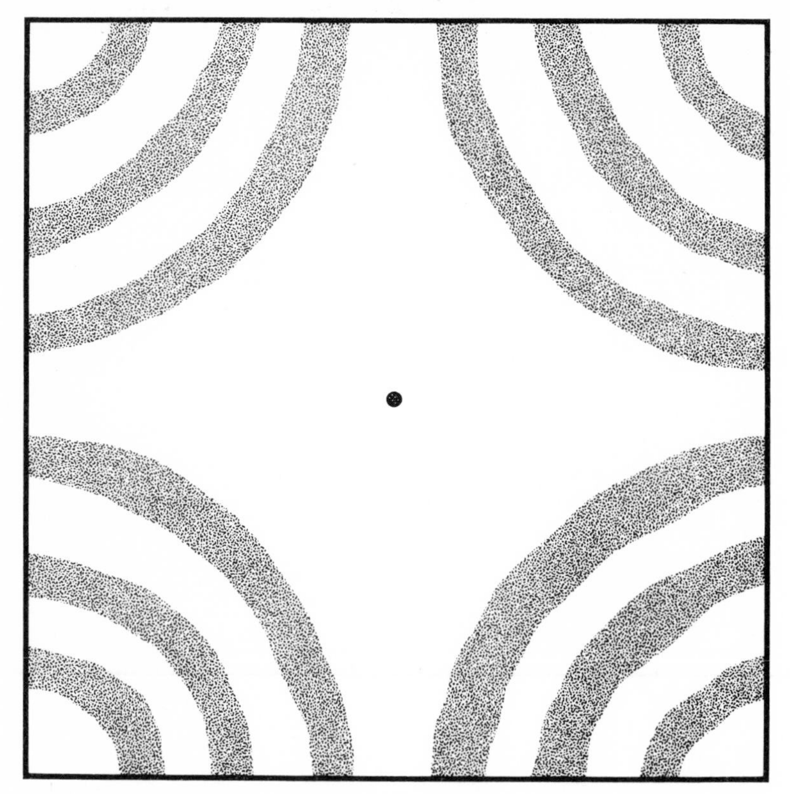

*Fig. 4.4* Patterns are formed in the sand at the resonant frequencies of a plate set in vibration using a vibrator controlled by an audio-frequency generator

$$f = 0{\cdot}45 \times 2350 \times 0{\cdot}12[(\tfrac{1}{4})^2 + (\tfrac{1}{2})^2]$$
$$= 0{\cdot}45 \times 2350 \times 0{\cdot}12 \times \tfrac{5}{16}$$
$$= 40 \text{ Hz}$$

## Coincidence Frequency

Under certain conditions the projected wavelength of incident sound is exactly equal to the length of resonant bending waves in the partition as shown in Fig. 4.5.

$$\lambda_B \sin\theta = \lambda$$

$$\therefore \quad \lambda_B = \frac{\lambda}{\sin\theta}$$

where $\lambda_B$ = bending wavelength
$\lambda$ = wavelength of sound in air
$\theta$ = angle of incidence

but $V_B = (1{\cdot}8\ \mathrm{hf}V_L)^{1/2}$

$$V_B = f\lambda_B$$
$$c = f\lambda$$

$$\therefore \quad V_B = \frac{\lambda_B \cdot c}{\lambda}$$

$$= \frac{c}{\sin\theta}$$

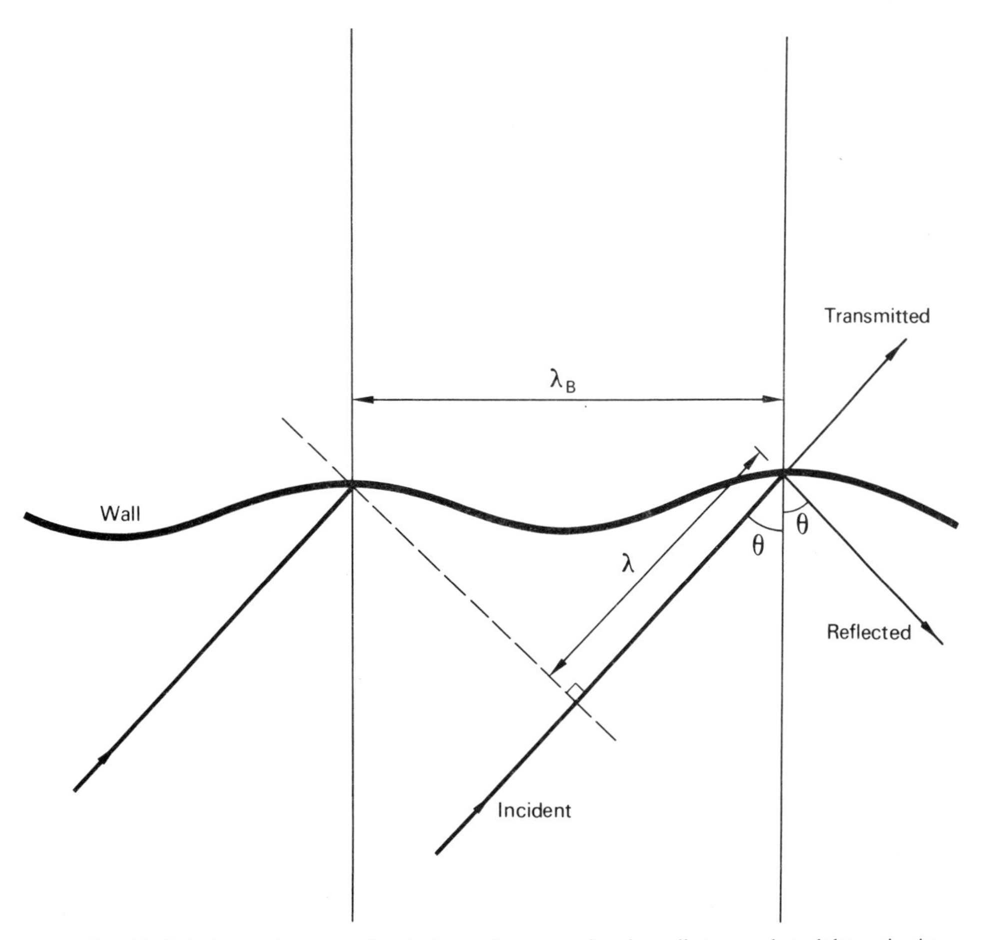


*Fig. 4.5* Coincidence frequency. Incident sound wave reaches the wall at an angle and the projection of the natural wavelength of vibration of the wall is equal to the sound wavelength

$$\therefore \quad \frac{c}{\sin\theta} = (1{\cdot}8\ hf\ V_L)^{1/2}$$

$$\therefore \quad f = \frac{c^2}{1{\cdot}8\ hV_L \sin^2\theta}$$

where $c$ = velocity of sound in air
$V_B$ = bending wave velocity
$h$ = thickness of panel
$V_L$ = longitudinal wave velocity

This frequency $f$ at which the effect is a maximum is known as the coincidence frequency. Because $f$ is inversely proportional to $V_L$, and $V_L$ is related to stiffness, the greater the stiffness the lower the coincidence frequency.

## Critical Frequency

This is the lowest frequency at which coincidence occurs.
Thus $\sin\theta = 1$

$$\therefore \quad \theta = 90°$$

$$f = \frac{c^2}{1{\cdot}8hV_L}$$

$$= \frac{c^2}{1{\cdot}8\,h\sqrt{\left(\frac{E}{\rho}\right)}}$$

$$= \frac{c^2}{1{\cdot}8\,h}\sqrt{\left(\frac{\rho}{E}\right)}$$

### Example 4·2

What is the expected critical frequency for a 120 mm thick brick wall? Assume a longitudinal wave velocity in brick of 2350 m/s and that the velocity of sound in air is 330 m/s.

$$f = \frac{330^2}{1{\cdot}8 \times 0{\cdot}120 \times 2350}$$

$$= 214{\cdot}5 \text{ Hz}$$

### Coupling of Source to the Structure

This is exactly similar to the mass law part of the preceding section except that the other media will not always be air. It is essential to obtain a high impedance mismatch between the source and the structure. In the case of machinery a resilient mount will normally be needed.

It can be seen from Fig. 4.6 that the natural frequency of the resilient mount must be much lower than the driving frequency of the machine. The resonant frequency of the mount can be found from its static deflection:

$$f = \sqrt{\left(\frac{250}{h}\right)}$$

where $h$ = static deflection in mm under the machine.

Deflections may be much greater in the case of non-elastic materials such as rubber ($\times 2$) and cork ($\times 4$).

### Example 4·3

A certain machine with a slightly out of balance motor rotating at 1800/min is fixed on a perfectly elastic mount with a static compression of 2·50 mm. Calculate the reduction in dB by use of the mounting.

Forcing frequency, $$F = \frac{1800}{60}$$

$$= 30 \text{ Hz}$$

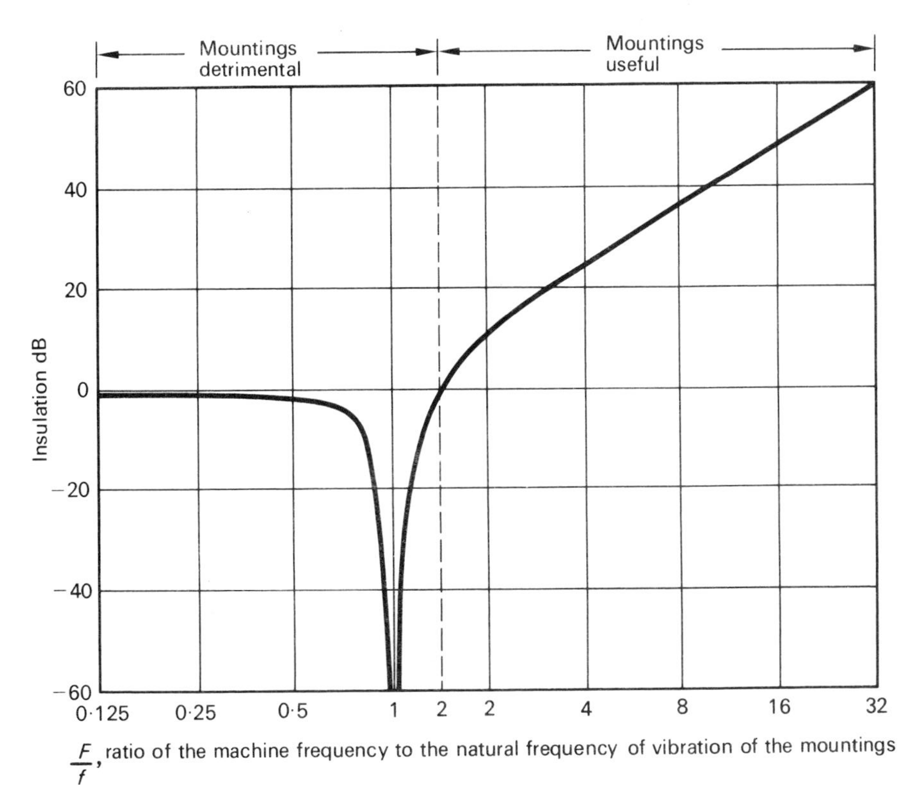


*Fig. 4.6* Energy reduction in dB by the use of resilient machine mountings

Resonant frequency of mount, $f$

$$= \sqrt{\left(\frac{250}{2 \cdot 50}\right)}$$

$$= \sqrt{100}$$

$$= 10 \text{ Hz}$$

$$\text{Ratio } \frac{F}{f} = \frac{30}{10}$$

$$= 3$$

$\therefore$ Insulation (from Fig. 4·6) = 17·5 dB

### Impact Sound Insulation

The most common cause of structure borne sound is that of footsteps, particularly in multi-storey dwellings. Clearly what is needed is a suitably large impedance between the source of impact and the radiating surfaces. This may be achieved by mass, by the use of resilient materials or a combination of both. Light-weight constructions are likely to cause problems because they need to be very stiff for structural reasons with a consequential loss in sound insulation.

## Measurement of Impact Sound Insulation

With impact sound it is the level of noise produced in the receiving room which is of interest and not the level of the impact sound itself. The standard method is to produce a constant level by impact with the source room floor. The level in the receiving room is measured and this is compared with levels known to be satisfactory (see Fig. 4.7).

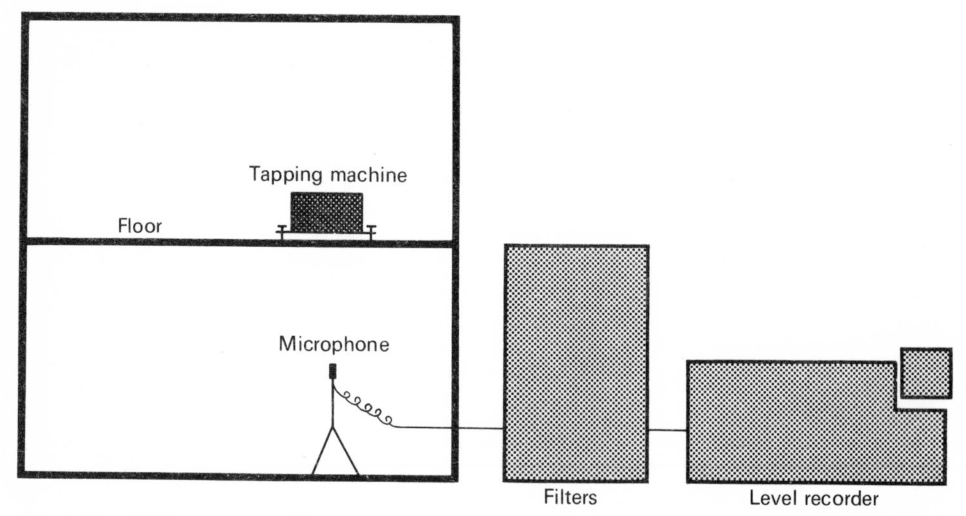


*Fig. 4.7* Measurement of impact sound insulation

The microphone and measuring apparatus are first calibrated at a known frequency by a pistonphone. This known level is marked on to paper using a level recorder. With the standard footsteps machine switched on, the levels in the lower room are measured in appropriate frequency intervals. These may be in octave, half or third octave bandwidths from 100 to 3200 Hz. (Octave or third octave bandwidth intervals are more common than half octaves.)

Usually six measurements will be made in each frequency band up to 500 Hz and three in each band above. The average should be calculated from

$$L_R = 10\log_{10}\frac{P_1^2+P_2^2+\ldots P_n^2}{nP_0}$$

where $P_0$ = reference sound pressure level
$= 2\times10^{-5}\ \mathrm{N/m^2}$

This is long and tedious, and provided that the maximum differences are less than 6 dB, it is sufficient to take the decibel average.

The results obtained in this way would be affected by the absorption of sound in the receiving room and not merely by the effectiveness of the floor construction. For comparison purposes the results are standardized to a fixed amount of room absorption. The room absorption is compared to 10 m$^2$ of absorption. This is the amount that would be present in the living room of an average dwelling.

$$\therefore\quad L_c = L_R - 10\log_{10}\frac{10}{A}$$

where $L_c$ = corrected level
$L_R$ = measured level
$A$ = area of absorption in the receiving room in $m^2$

If other than octave measurements are made these will have to be adjusted to the octave equivalents. One third octave results will need an addition of 5 dB. Measurements are made with one third of the energy.

$$\therefore \quad \text{correction in dB} = 10 \log_{10} \tfrac{3}{1} = 10 \times 0{\cdot}4771 = 4{\cdot}771 \simeq 5\ \text{dB}$$

For half octave measurements correction in dB

$$= 10 \log_{10} 2 = 10 \times 0{\cdot}3010 = 3\ \text{dB}$$

In Britain, the correction to allow for the reverberation time of the receiving room is slightly different for domestic dwellings. It has been found that most dwelling rooms have a reverberation time very close to 0·5 s. For simplicity, insulation standards between domestic dwellings are corrected to this 0·5 s reverberation time.

$$\text{Hence: } L_c = L_R - 10 \log_{10} \frac{t}{0{\cdot}5}$$

Where $t$ = actual reverberation time of the receiving room in s.

On the assumption that Sabine's formula is correct, the two corrections are identical.

$$\frac{t}{0{\cdot}5} = \frac{\dfrac{0{\cdot}16\,V}{A}}{\dfrac{0{\cdot}16\,V}{10}}$$

$$\frac{t}{0{\cdot}5} = \frac{10}{A}$$

Departures from Sabine's formula explained in the previous chapter will lead to minor discrepancies in the correction term.

### Impact Sound Insulation Standards

There are two grade curves in Britain (see Fig. 4.8). Grade 1 gives the highest economic standard that is possible for flats. It should cause little or no complaint. Grade 2 is likely to be satisfactory for many people — particularly if there is plenty of masking noise from television, road traffic, etc. There will often be considerable dissatisfaction which must be offset against greater economy in rent. Many constructions will be satisfactory in most frequency bands but fall short in a few.

To comply with a particular grade, the sum of the failures for one third octave measurements must not be greater than 23 dB. In general the average difference must

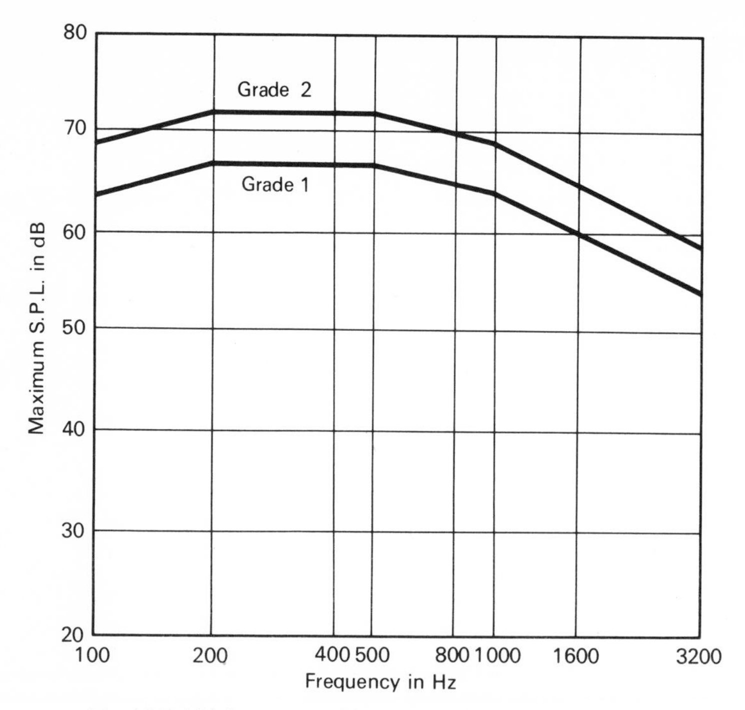


*Fig. 4.8* British impact sound insulation grade curves

be less than 1·5 dB. It should be noticed that only the failures are taken into account, and very high insulation at some frequencies cannot be used to offset poor insulation at others.

The international standard is similar except that only one grade curve exists and an insulation index is computed (see Fig. 4.9). This is done by shifting the standard curve in 1 dB intervals until either:

1. the mean unfavourable deviation is less than 2 dB and the maximum does not exceed 8 dB for one third octaves or 5 dB for octaves, or
2. the mean unfavourable deviation is between 1 and 2 dB.

Whichever is the more severe is used. The insulation index is then the value of the shifted reference curve at 500 Hz.

### Example 4·4

The mean corrected levels in the receiving room found from a measurement of impact sound insulation were:

| Hz | 100 | 125 | 160 | 200 | 250 | 315 | 400 | 500 |
|---|---|---|---|---|---|---|---|---|
| dB | 69 | 68 | 67 | 70 | 70 | 70 | 72 | 71 |
| Hz | 630 | 800 | 1000 | 1250 | 1600 | 2000 | 2500 | 3150 |
| dB | 67 | 67 | 61 | 58 | 53 | 50 | 48 | 46 |

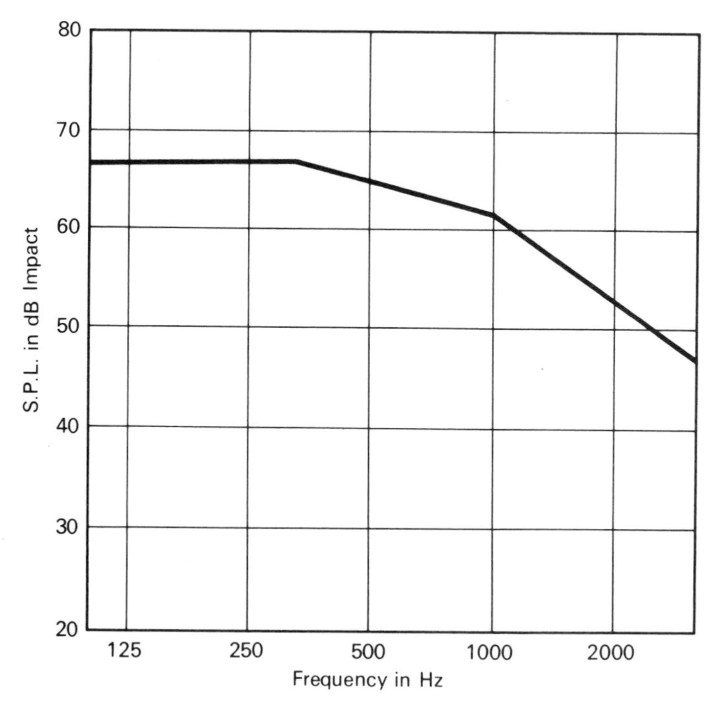


*Fig. 4.9* ISO impact sound grade curve

Compare the insulation with the British Grade 2 Standard and find the ISO insulation index. (Assume no floor covering.)

| Frequency Hz | S.P.L. in dB | Grade 2 | Deviation from Grade 2 | Grade 1 | Deviation from Grade 1 | ISO St. | Deviation from ISO |
|---|---|---|---|---|---|---|---|
| 100 | 69 | 69 | 0 | 64 | −5 | 67 | −2 |
| 125 | 68 | 70 | | 65 | −3 | 67 | −1 |
| 160 | 67 | 71 | | 66 | −1 | 67 | |
| 200 | 70 | 72 | | 67 | −3 | 67 | −3 |
| 250 | 70 | 72 | | 67 | −3 | 67 | −3 |
| 315 | 70 | 72 | | 67 | −3 | 67 | −3 |
| 400 | 72 | 72 | | 67 | −5 | 66 | −3 |
| 500 | 71 | 72 | | 67 | −4 | 65 | −6 |
| 630 | 67 | 71 | | 66 | −1 | 64 | −3 |

| Frequency Hz | S.P.L. in dB | Grade 2 | Deviation from Grade 2 | Grade 1 | Deviation from Grade 1 | ISO St. | Deviation from ISO |
|---|---|---|---|---|---|---|---|
| 800 | 67 | 70 | | 65 | −2 | 63 | −4 |
| 1000 | 61 | 69 | | 64 | | 62 | |
| 1250 | 58 | 67 | | 62 | | 59 | |
| 1600 | 53 | 65 | | 60 | | 56 | |
| 2000 | 50 | 63 | | 58 | | 53 | |
| 2500 | 48 | 61 | | 56 | | 50 | |
| 3150 | 46 | 59 | | 54 | | 47 | |
| | | | | | −30 | | −31 |

The particular floor does not conform to the British Grade 1 floor construction because the maximum deviation exceeds 23 dB. It is therefore Grade 2.

$$\text{ISO mean unfavourable deviation} = \frac{+31}{16}$$

$$= 1\frac{15}{16}\text{ dB}$$

ISO insulation index is therefore 0.

### Building Regulations

Section G of the British Building Regulations (Parts 3 and 4) cover the types of floor construction ‘deemed to satisfy’ between different dwellings. It will be noticed that all these constructions conform to Grade 1 Standards if there is any chance of structure borne noise (e.g. machinery and tank rooms), but Grade 2 is permissible in less demanding circumstances.

### Questions

(1) An elastic mounting for a vibrating machine has a static deformation of 1·13 mm and gives an energy reduction of 9 dB when the machine is running steadily. Calculate the forcing frequency of the machine.

(2) An insulation of 40 dB is required by the use of an elastic mount for a machine vibrating at 2400 rev/min. What static deflection would be needed?

(3) A 50 mm thick concrete panel supported at its edges was found to have a lowest resonant condition at 56 Hz. An ultrasonic pulse was found to take 10 $\mu$s to travel through the thickness of the concrete. Find the dimensions of the panel.

(4) A 13 mm thick plasterboard has a critical frequency of 2000 Hz and a density of 1700 Kg/m$^3$. Calculate Young’s modulus of elasticity. ($V = 333$ m/s.)

(5) Third octave impact insulation measurements were found to give the following results for a particular floor:

| Frequency Hz | Mean level in dB re $2 \times 10^{-5}$ N/m$^2$ | R.T. of receiving room in s |
|---|---|---|
| 100 | 76 | 2·35 |
| 125 | 75 | 1·95 |
| 160 | 70 | 2·2 |
| 200 | 65 | 1·95 |
| 250 | 62 | 1·5 |
| 315 | 59 | 2·1 |
| 400 | 60 | 1·8 |
| 500 | 57 | 1·8 |
| 630 | 50 | 1·4 |
| 800 | 49 | 1·1 |
| 1000 | 49 | 1·1 |
| 1250 | 49 | 1·1 |
| 1600 | 42 | 1·0 |
| 2000 | 40 | 1·0 |
| 2500 | 38 | 0·8 |
| 3150 | 35 | 0·75 |

(a) State whether these results would make the construction satisfactory British Grade 1 or Grade 2 Standards for impact sound insulation and calculate the mean difference.
(b) Find the ISO impact insulation index. (It may be assumed that the Sabine formula is correct.)

(6) The following results were obtained for the impact sound insulation of a floor, together with values for Grade 2 insulation. On checking the calibration it was found that a reading of 91 dB was obtained instead of 90 dB. Compare the results graphically with the standard given and state if, and by how much, the result fails it.

| Frequency Hz | 100 | 200 | 400 | 800 | 1600 | 3200 |
|---|---|---|---|---|---|---|
| Grade 2 dB | 68 | 72 | 72 | 70 | 65 | 58 |
| Result dB | 73 | 75 | 73 | 67 | 56 | 46 |
| R.T. of receiving room in s | 1·5 | 1·4 | 1·4 | 1·3 | 1·3 | 1·2 |

(7) Find the velocity of sound in a concrete beam of density 2400 kg/m$^3$ and with a dynamic modulus of elasticity ($E$) of 38 400 N/mm$^2$.

Chapter 5

# Airborne Sound

The term airborne sound is conveniently taken to mean sound that is transmitted mainly, but not necessarily exclusively, through the air. For instance, sound transmitted from one room to another may have to excite the adjoining wall into vibration. This chapter looks at the propagation of airborne sound and the methods which may be used to reduce it.

## Propagation of Sound in Open Air

### Point Source

It was shown in Chapter 1 that air cannot sustain a shear force and only a longitudinal type of sound wave is possible. Sound will radiate spherically in a free field so that the sound intensity will decrease with the square of the distance from the source (see Fig. 5.1).

$$\therefore \quad I \text{ is proportional to } \frac{1}{d^2}$$

$$\therefore \quad \frac{I_r}{I_R} = \frac{R^2}{r^2}$$

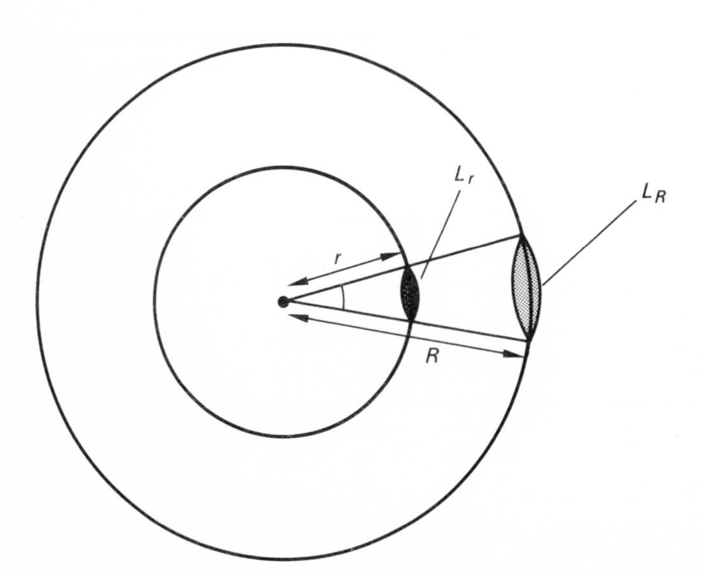


*Fig. 5.1* The surface area of a sphere is proportioned to the square of its radius. The sound intensity will therefore decrease in proportion to the square of the distance from the point source

where $I_r$ = intensity at a distance $r$ from the source

$I_R$ = intensity at a distance $R$ from the source

If $L_r$ = sound pressure level in decibels at distance $r$ from the source

$$\text{Then } L_r - L_R = 10 \log_{10} \frac{I_r}{I_R}$$

$$= 10 \log_{10} \frac{R^2}{r^2}$$

$$= 20 \log_{10} \frac{R}{r}$$

Thus the reduction in sound pressure level for sound from a point source will be 6 dB for each time the distance is doubled.

$$L_r - L_{2r} = 20 \log_{10} \frac{2r}{r}$$

$$= 20 \times 0{\cdot}3010$$

$$\simeq 6 \text{ dB}$$

If the sound power level in picowatts ($L_w$) had been given, then the formula becomes:

$$L_r - L_w = 20 \log_{10} \frac{1}{r} - 10{\cdot}9 \text{ (} r \text{ in metres)}$$

or $\quad L_r = L_w - 20 \log_{10} r - 10{\cdot}9$

If the sound field is confined to a hemisphere because the source is on the (perfectly reflecting) ground then the intensity is doubled or the sound pressure level is increased by 3 dB.

$$\therefore \quad L_r = L_w - 20 \log_{10} r - 7{\cdot}9$$

**Example 5·1**

The sound power level of a certain jet plane flying at a height of 1 km is 160 dB (re $10^{-12}$ $W$). Find the maximum sound pressure level on the ground directly below the flight path assuming that the aircraft radiates sound equally in all directions.

$$\begin{aligned} L_r &= 160 - 20 \log_{10} 1000 - 7{\cdot}9 \\ &= 160 - 20 \times 3 - 7{\cdot}9 \\ &= 160 - 60 - 7{\cdot}9 \\ &= 92{\cdot}1 \text{ dB} \end{aligned}$$

It should be noted that the assumption that sound is radiated equally in all directions is not correct although the figures given are of the correct order.

LINE SOURCE

When the sound source is a line instead of a point, such as is the case near a busy motorway, the sound will radiate in the form of a cylinder (Fig. 5.2).

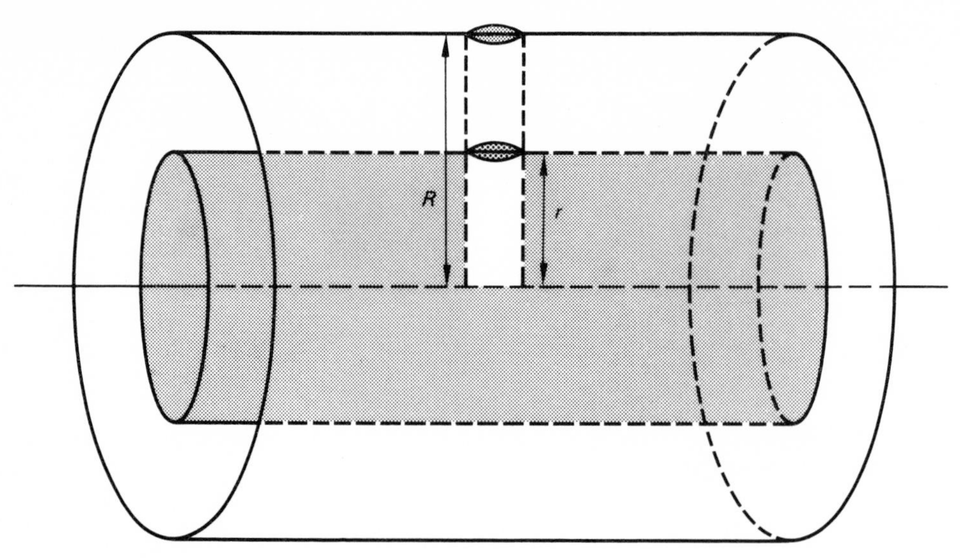


*Fig. 5.2* The surface area of a cylinder is proportional to its radius. Sound intensity will therefore decrease directly with distance from a line source

Thus $$\frac{I_r}{I_R} = \frac{R}{r}$$

$$\therefore \quad L_r - L_R = 10 \log_{10} \frac{R}{r}$$

If $R = 2r$

Then $$L_r - L_R = 10 \log_{10} 2 = 10 \times 0{\cdot}3010 = 3 \text{ dB}$$

It can be seen that there is only a 3 dB reduction for a doubling of the distance.

### Example 5·2

A policeman measures the sound pressure level at a distance of 7·5 m from the line of traffic on a road and finds it to be 80 dB. What would the level be at a distance of 75 m from the line of traffic if the policeman's reading was from:

(a) an isolated vehicle?

(b) a continuous line of closely packed identical cars? (Assume that the ground is flat and unobstructed.)

(a) The isolated vehicle constitutes a point source

$$\therefore \quad \text{reduction in dB} = 10 \log_{10} \left[\frac{75}{7{\cdot}5}\right]^2$$

$$= 20 \log_{10} 10$$

$$= 20 \times 1$$

$$= 20 \text{ dB}$$

The sound pressure level at 75 m from the vehicle would therefore be $80 - 20 = 60$ dB.
(b) The continuous line of closely packed cars could be taken as a line source.

$$\therefore \quad \text{reduction in dB} = 10 \log_{10} \frac{75}{7{\cdot}5}$$

$$= 10 \log_{10} 10$$

$$= 10 \text{ dB}$$

The level at 75 m from the line of cars would be

$80 - 10 = 70$ dB

## Reduction of Noise by Walls

The reduction of sound by means of a wall or fence is only effective where the barrier is large compared with the wavelength of the noise. It has been shown that approximately:

$$x = \frac{H^2}{\lambda D_S}$$

Where $x$, $H$ and $D_S$ are shown in Fig. 5.3, and where $x$ is related to the sound reduction in dB as shown in Fig. 5.4. Provided that $D_L \gg D_S$ and $D_S > H$.

## Example 5·3

A school is situated close to a furniture factory with a flat roof on top of which is a dust extractor plant as shown in Fig. 5.5. The highest window in the school facing the factory is at the same height as the noise source, which can be taken as 1 m above the factory roof. The extractor plant is 6 m from the edge and produces a note of 660 Hz. Find the minimum height of wall to be built on the edge of the factory roof to give a reduction of 15 dB.

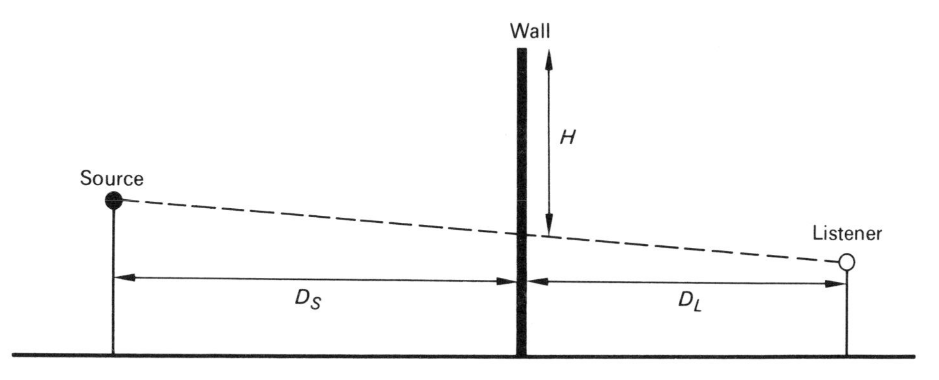


*Fig.* 5.3 Relation of source and listener to wall

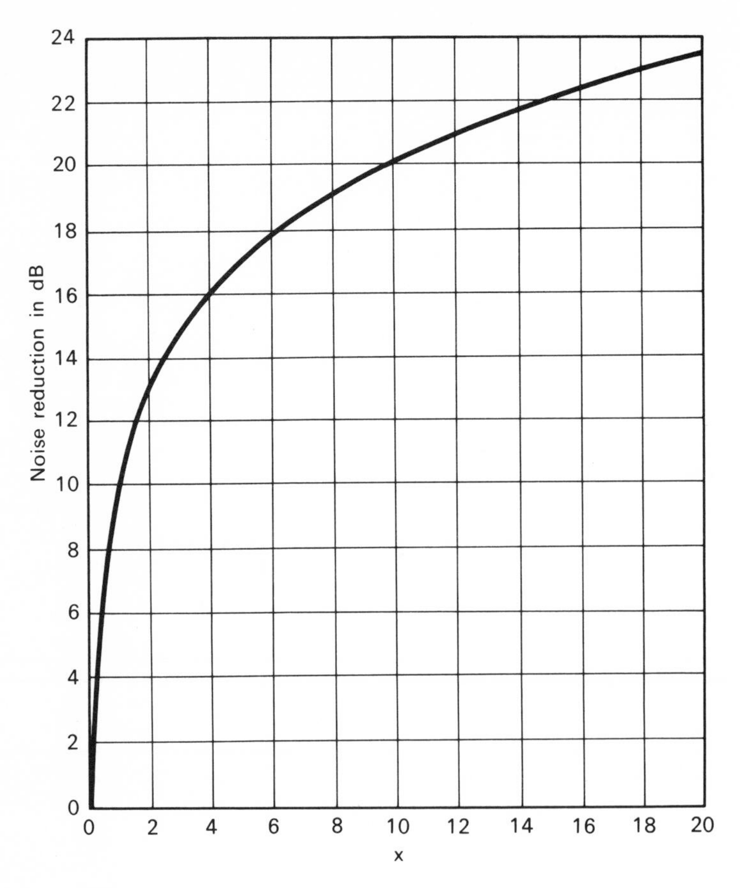


*Fig. 5.4* Sound reduction by means of a wall or screen

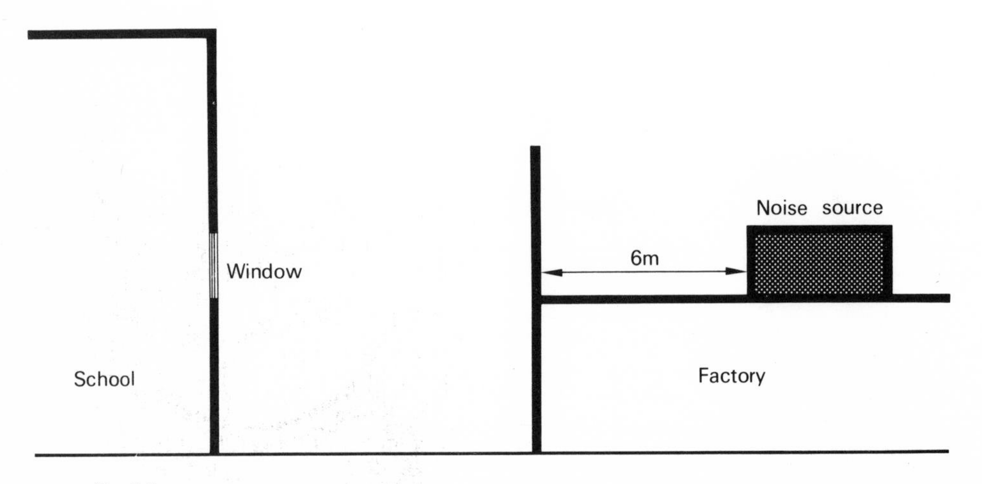


*Fig. 5.5*

From Fig. 5.4, the value of $x$ required to give 15 dB reduction is 3

$$\therefore \quad 3 = \frac{H^2}{V/f \,.\, D_S}$$

$$= \frac{H^2}{\frac{330}{660} \,.\, 6}$$

$$= \frac{H^2}{3}$$

$$\therefore \quad H^2 = 9$$

$$H = 3 \text{ metres}$$

Total height of wall

$$= 4 \text{ metres}$$

If the distance $D_L$ from the wall to the listener is not large compared with $D_S$

$$\text{then } x = \frac{2}{\lambda}\left[D_s\left(\sqrt{\left\{1+\frac{H^2}{D_s^2}\right\}}-1\right)+D_L\left(\sqrt{\left\{1+\frac{H^2}{D_L^2}\right\}}-1\right)\right]$$

These calculations are based on the amount of sound diffracted around the barrier. It is essential that the barrier itself does not transmit sound. In effect this means that the insulation of the barrier must be greater than the screening effect needed. Gaps in a fence would make the screen useless although it can be quite light in weight. A mass of 20 kg/m$^2$ is usually sufficient.

### Bushes and Trees

The noise attenuation achieved by shrubs and trees can only be marginal. Measurements made in jungle conditions allowing for the normal loss by distance alone are shown approximately in Fig. 5.6.

Considering the marginal improvement, trees are not an economic means of achieving sound insulation. However, there is some absorption effect which would not be present if the ground were paved. If trees are used they must be of a leafy variety and be thick right to ground level. A point sometimes forgotten is that the rustle of leaves may occasionally produce a noise level as high as 50 dBA.

### Air Absorption

Energy is absorbed as a sound wave is propagated through the air. These losses are due to a relaxation process and depend upon the amount of water vapour present. They are approximately proportional to the square of the frequency.

The attenuation per metre, $\alpha$, has been shown to be such that:

$$\alpha = kf^2 + \alpha_2$$

where $k = 14{\cdot}24 \times 10^{-11}$

$f$ = frequency in Hz

$\alpha_2$ is humidity dependent.

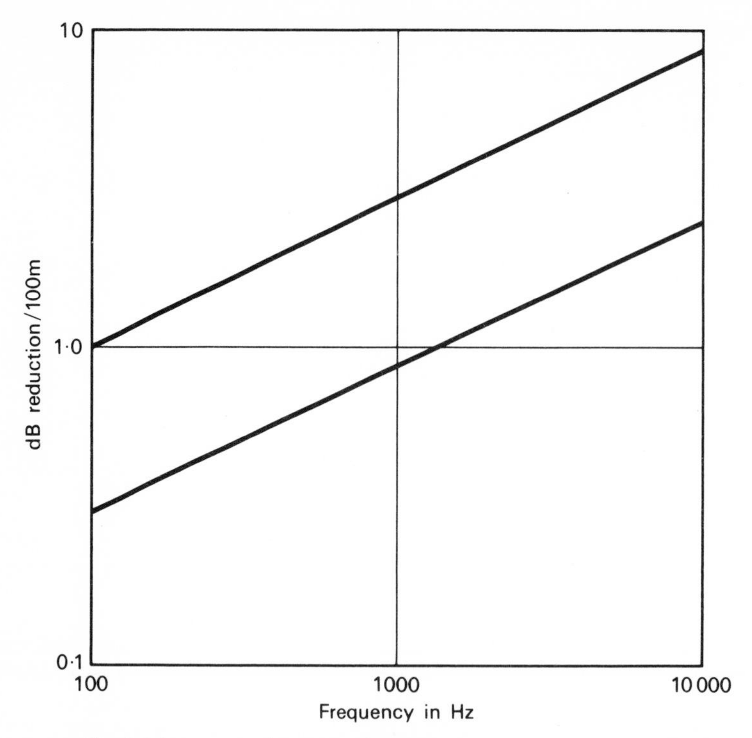


*Fig. 5.6* Attenuation by means of trees. The upper line represents maximum attenuation under conditions of very dense growth of leafy trees and bushes where penetration is only possible by cutting. The lower line shows the attenuation under conditions where penetration is easy and visibility up to 100 m, but still very leafy undergrowth

Typical values are about 3 dB per 100 metres at 4000 Hz dropping to 0·3 dB per 100 metres at 1000 Hz. Air attenuation becomes very important for ultrasonic frequencies and is greater than 1 dB/m at 100 kHz, but is of comparatively little significance in architectural acoustics (see Fig. 5.7).

## Velocity Gradients

For propagation close to the ground, sound velocity gradients have a big influence on the levels received from a distance. These velocity gradients can be caused by wind or temperature.

Friction between the moving air and the ground results in a decreased velocity near ground level. This causes a distortion of the wave front. Downwind from the source, sound rays are refracted back towards the ground and the received level is not affected. Upwind the sound is refracted up and away from the ground causing acoustic shadows in which the level is considerably reduced (Fig. 5.8).

The velocity of sound in air is proportional to the square root of the absolute temperature. A temperature lapse condition will cause the sound rays to be bent up causing a symmetrical shadow around the source (see Fig. 5.9).

A motorway built downward (east) of a town would have an advantage of about 10 dB extra attenuation over one to the west.

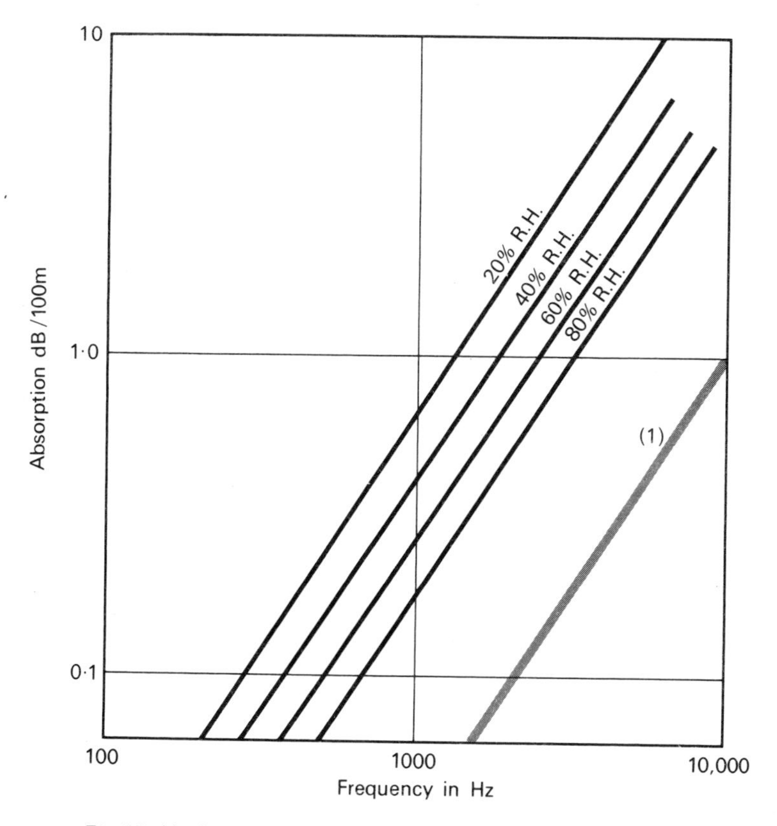


*Fig. 5.7* Air absorption dB/100 m. Curve (1) shows the absorption due to the $kf^2$ part of the equation. The other lines show the total contributions

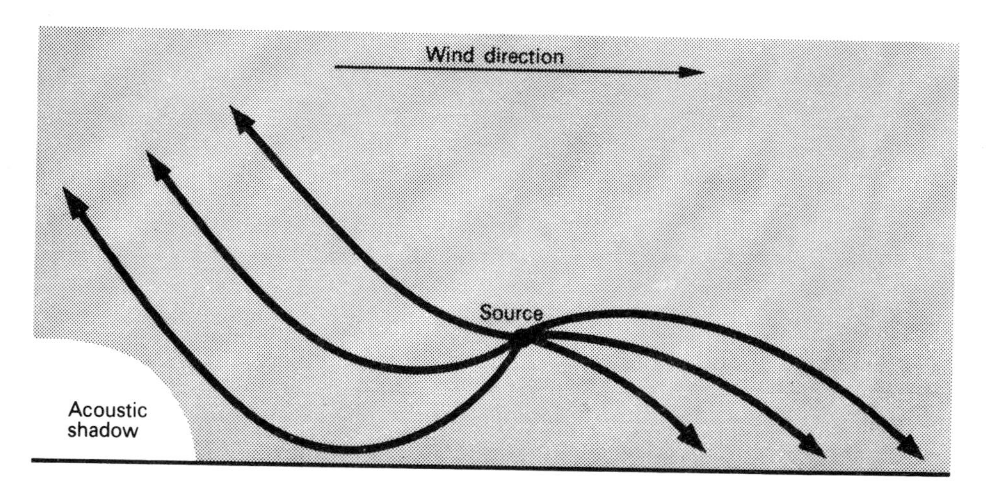


*Fig. 5.8* Wind-created velocity gradients

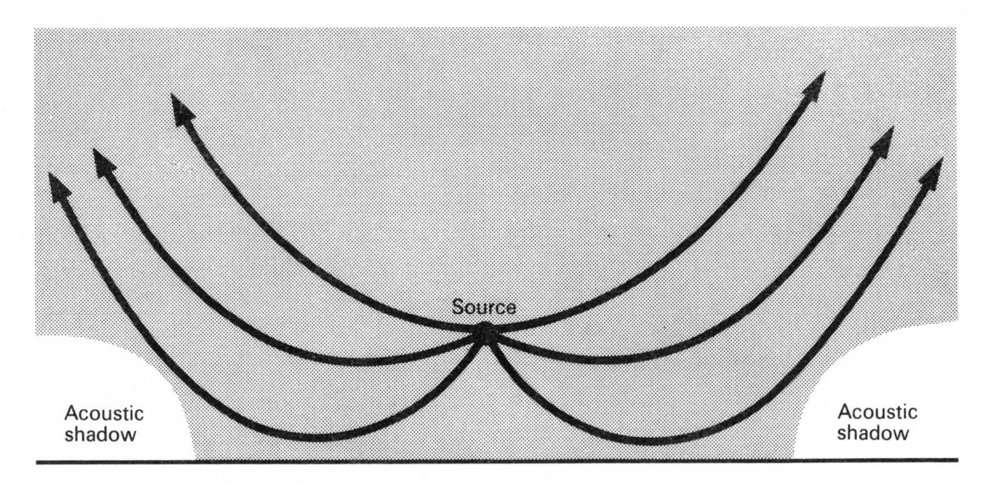


*Fig. 5.9* Temperature lapse-created velocity gradients

## Ground Absorption

Hard surfaces such as concrete appear to have virtually no absorption properties, whereas there may be a maximum attenuation over grassland of 25 to 30 dB in the 200 to 400 Hz octave band along a distance of about 1000 metres. It seems that there is a preferential absorption in this frequency range for sound produced near the ground.

## Sound Insulation by Partitions

Sound can be transmitted into a room by some or all of the methods shown in Fig. 5.10.

1. Airborne sound in the source room excites the separating partition into vibration which directly radiates the sound into the receiving room. The amount of attenuation will depend upon the frequency of sound, the mass, fixing conditions and thus the resonant frequencies of the partition.
2. Airborne sound in the source room may excite walls other than the separating one into vibration. The energy is then transmitted through the structure and reradiated by some other partition into the receiving room.
3. Any wall other than the separating one may be excited. The sound is transmitted to the separating wall and then reradiated by it.
4. Sound energy from the separating partition is radiated into the receiving room by some other wall.

This means that a laboratory measurement of the sound insulation provided by a partition, which is mounted in massive side walls, may give results quite different from those achieved in actual buildings. There is a limit to the insulation obtained by improving only the adjoining partition. Where a partition has a low insulation value of 35 dB or less, flanking transmission is of little consequence, but when partition values of 50 dB are reached, further improvement is limited by the indirect sound paths.

The sound transmission properties of a partition can be divided into three distinct regions (see Fig. 5.11).

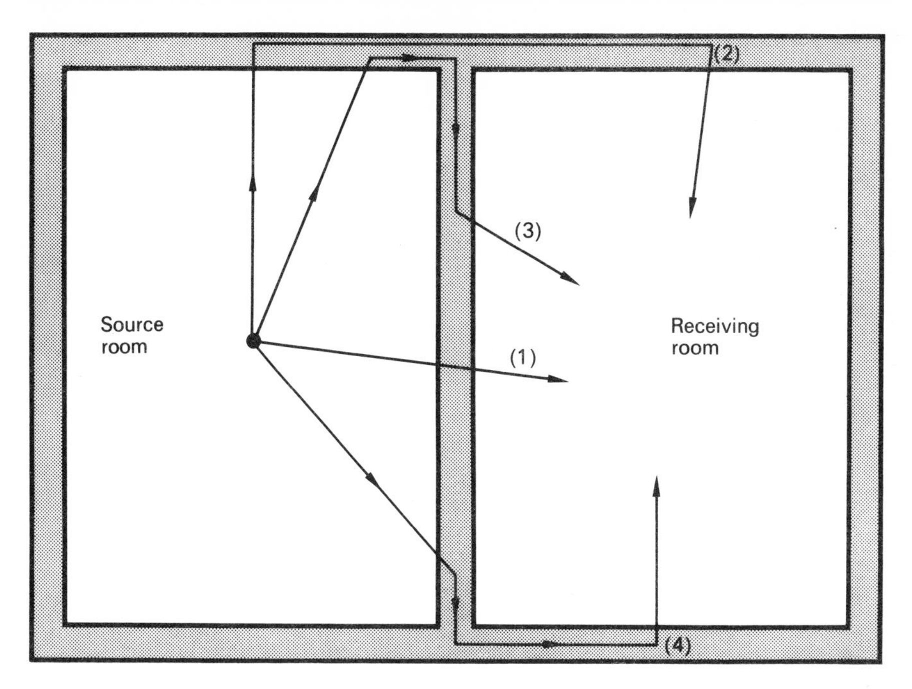


*Fig. 5.10* Four different transmission paths from source room to receiving room

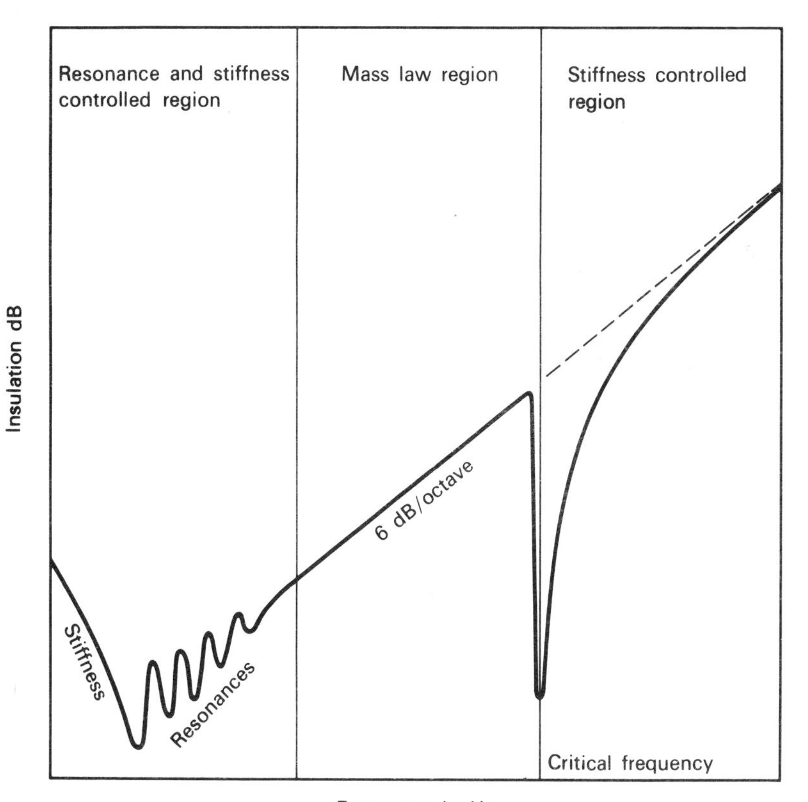


*Fig. 5.11* Three distinct regions showing the way a panel will react to different frequency sounds

Region 1 where resonance and stiffness conditions are important.

Region 2 which is mass controlled and where the partition can be considered as a large number of small masses free to slide over each other.

Region 3 where again the partition becomes stiffness controlled above a certain frequency known as the critical frequency.

Most building materials fall into the middle category. Typical values of coincidence frequency can be obtained from Table 5·1. A 50 mm reinforced concrete panel of mass per unit area 120 kg/m$^2$ can be expected to have a coincidence frequency at about 800 Hz.

**Table 5·1**

| Material | Mass/unit area × coincidence frequency × $10^3$ kg/m$^2$ × Hz |
|---|---|
| Plywood | 17 |
| Glass | 35 |
| Plasterboard | 50 |
| Concrete | 100 |
| Steel | 150 |

### Measurement of Airborne Sound Insulation of Panels

The field measurement consists of producing a suitable sound on one side of the panel and measuring the reduction in sound pressure level at the other side. Ideally, a completely diffuse sound field is needed in the source room. To attempt to eliminate the variations due to room modes for each frequency a band of noise is used, either in the form of white noise or in warble tones. A warble tone is a sound whose frequency is continuously varying in a regular manner within fixed limits.

If a white noise is used then the measurements may be made in one third, half or octave intervals, the centre frequencies being:

For one third octave:

100, 125, 160, 200, 250, 315, 400, 500, 640, 800, 1000, 1250, 1600, 2000, 2500, 3150 Hz

For half octave or octave:

Starts at 100 Hz and goes up to 3150 Hz
or at 125 Hz and goes up to 4000 Hz

If warble tones are used then the frequency deviation should be at least $\pm 10$ per cent of the mean frequency with a modulation of around 6 Hz. Above 500 Hz a deviation of $\pm 50$ Hz is enough.

Suitable means of producing the sound can be by the use of a random noise generator for white noise or a beat frequency oscillator for warble tones.

The measurement is made by finding the average sound pressure level difference between source and receiving rooms. This may conveniently be done by using two microphones (whose calibration has been checked) connected to a switchbox through filters to a level recorder as shown in Fig. 5.12. By this means the difference may be read off directly. The average is found over the entire room excepting those parts where direct radiation or boundary reflection are of significance. The average level, $L$, should be taken as:

$$L = 10 \log_{10} \frac{P_1^2 + P_2^2 + P_3^2 + \ldots P_n^2}{nP_0^2}$$

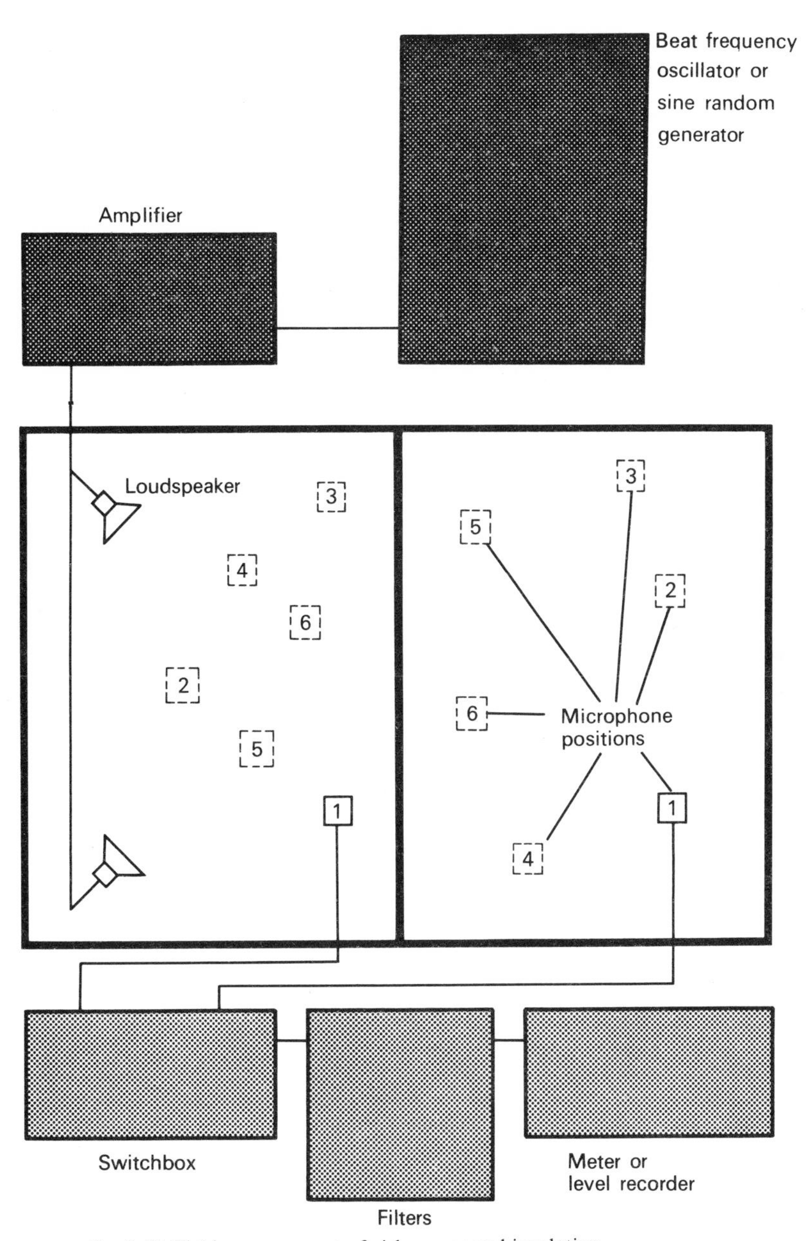


*Fig. 5.12* Field measurement of airborne sound insulation

where $P_1, P_2 \ldots P_n$ = RMS levels at $n$ different points in the room
$P_0$ = reference sound pressure

However, in practice so long as the values of $P_1, P_2, \ldots P_n$ do not differ greatly it is satisfactory to find the average level, $L$, from:

$$L = \frac{L_1 + L_2 + L_3 + L_4 + \ldots L_n}{n}$$

where $L_1, L_2, L_3 \ldots L_n$ = measured pressure levels at $n$ different points in the room

The average sound pressure level difference, $D$, is then found from:

$$D = L_S - L_R$$

where $L_S$ = average sound pressure level in the source room
$L_R$ = average sound pressure level in the receiving room

Six different positions are usually sufficient in each room up to 500 Hz and three each above this frequency. If the measured levels at any frequency in either room differ by more than 6 dB then more measurements are needed so that the dB average does not differ from the pressure average by more than $\pm 1$ dB.

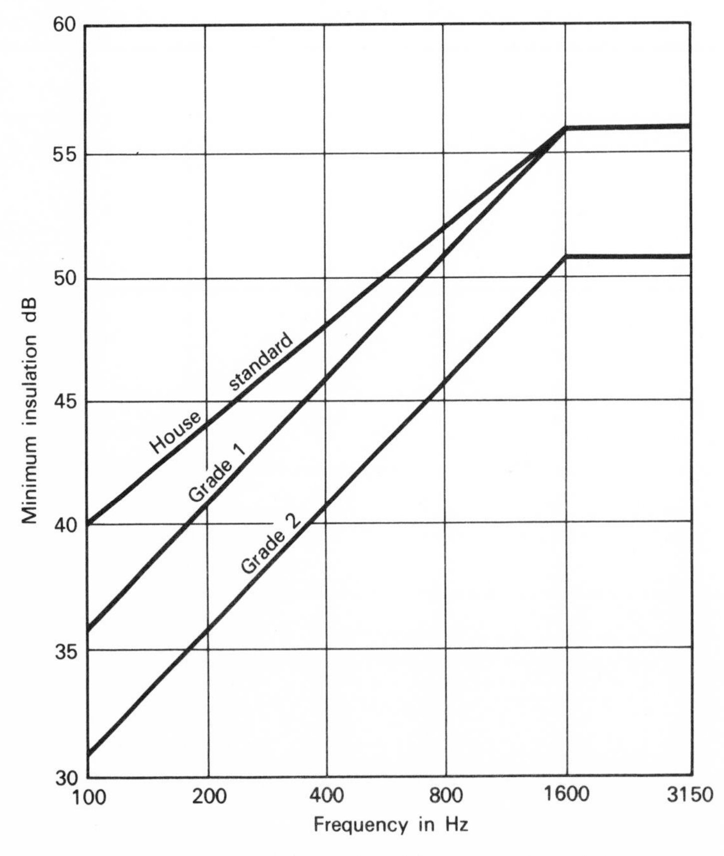


*Fig. 5.13* British grade curves for airborne sound insulation between dwellings

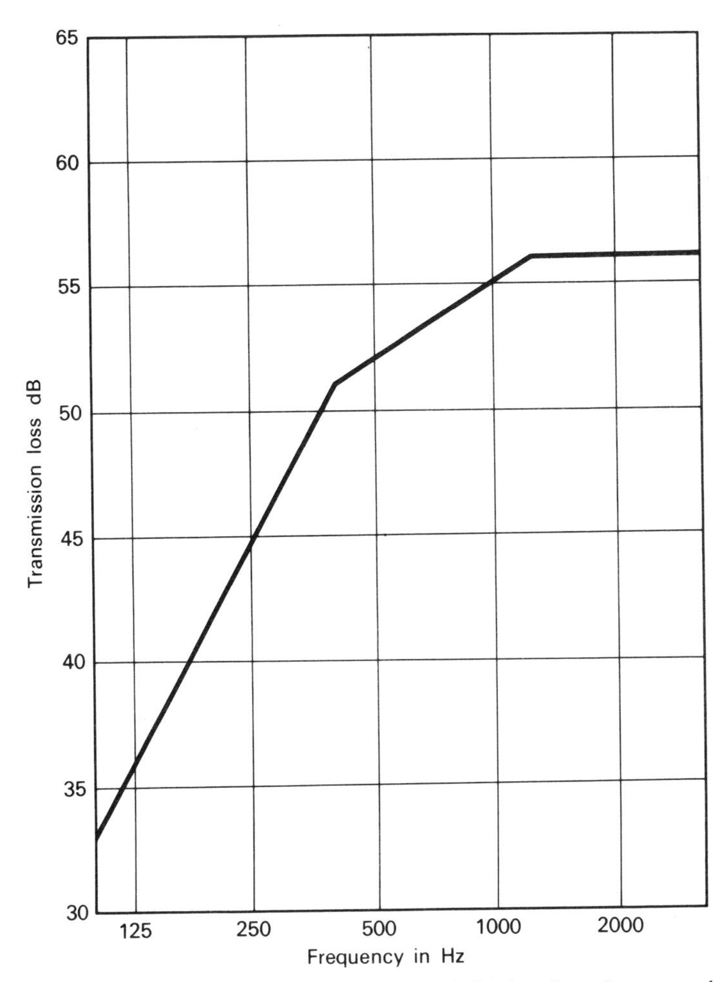


*Fig. 5.14* ISO reference values of transmission loss for airborne sound between dwellings

Variations in results for a particular partition could be produced by changes in the reverberation time of the receiving room. It is necessary for comparison purposes to normalize the result to a particular standard. The standard method adjusts the value to that which would have been obtained if the receiving room had 10 m$^2$ of absorption units. Hence:

$$D_n = L_S - L_R + 10 \log_{10} \frac{10}{A}$$

where $D_n$ = normalized level difference
$A$ = measured absorption in the receiving room in m$^2$

In Britain it has been found that for domestic dwellings the reverberation time seldom varies much from 0·5 s. A slightly different method of normalizing is used:

$$D_n = L_S - L_R + 10 \log_{10} \frac{t}{0{\cdot}5}$$

Where $t$ = reverberation time of the receiving room in seconds.

### Airborne Insulation Standards

Allowing for the fact that the method of normalizing is slightly different, the ISO reference values are very similar to the Grade 1 Standard for sound insulation (for flats) used in Britain (see Figs. 5.13 and 5.14). The reason for three curves in Britain is based on the fact that the house standard can be achieved by the horizontal attenuation provided by a 230 mm brick party wall, which is considered the desirable minimum. In flats the economic limit for airborne insulation is set by the vertical attenuation provided by a concrete construction with a floating floor. This reaches Grade 1 Standard for flats. The Grade 2 curve allows for more economic construction but would lead to some occupants being disturbed by noise from neighbours. The building regulations specify a minimum, equivalent to Grade 1, between dwellings.

Some examples of party wall and floor constructions are given in Table 5·2 and Fig. 5.15. Average sound insulation values, while simple, are of limited value and a frequency related insulation curve is preferable.

**Table 5·2**

*Grading of Party Walls*

| Construction | Weight kg/m$^2$ | Grade |
|---|---|---|
| 230 mm brick wall (solid) | 415 | house standard |
| 280 mm cavity brick wall | 415 | house standard |
| Concrete, panels or in situ | 415 | house standard |
| Lightweight aggregate with sound absorbent surfaces to: | | |
| (a) cavity of 50 mm | 300 | house standard |
| (b) cavity of 75 mm | 250 | house standard |
| (c) cavity of 50 mm | 250 | Grade 1 |

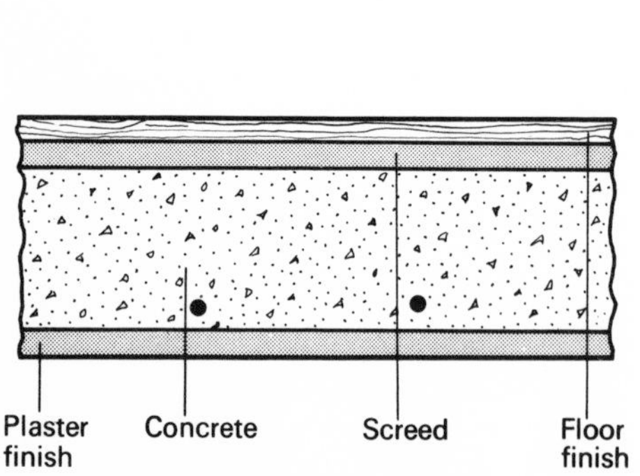


*Fig. 5.15(a)* This gives *GRADE 1* insulation for impact and airborne sound if: concrete + plaster + screed weighs more than 365 kg/m$^2$ and a soft floor finish is used. If a hard floor finish is used airborne insulation is still Grade 1, but the floor fails Grade 2 for impact sound

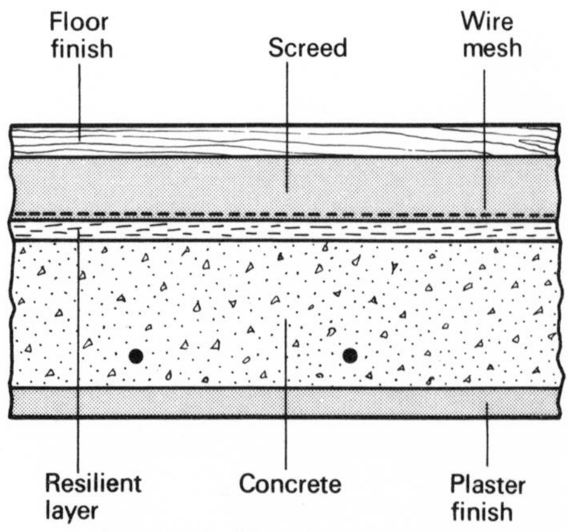


*Fig. 5.15(b)* This gives *GRADE 1* insulation for airborne and impact sound if: concrete + plaster weigh more than 220 kg/m$^2$

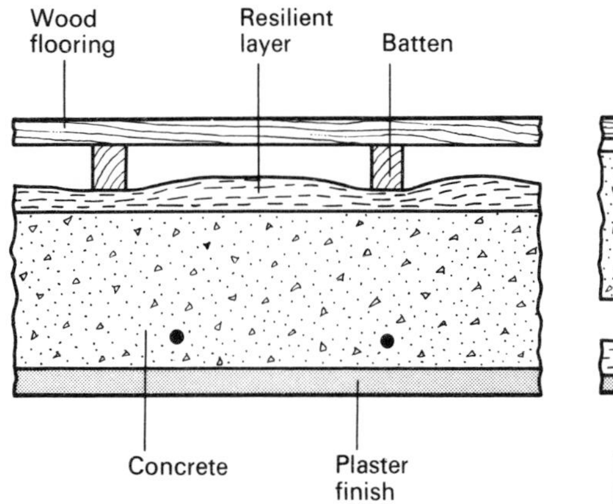


*Fig. 5.15(c)* This gives *GRADE 1* insulation for airborne and impact sound if: concrete and plaster weigh more than 220 kg/m$^2$

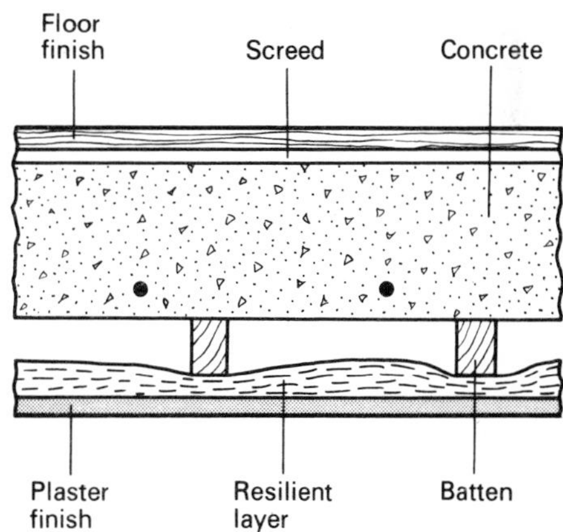


*Fig. 5.15(d)* This gives *GRADE 1* insulation for airborne and impact sound if: concrete + screed weigh more than 220 kg/m$^2$ and a soft floor finish is used. If a medium floor finish is used airborne insulation is still Grade 1 but the floor is only Grade 2 for impact sound. If a hard floor finish is used airborne insulation is still Grade 1, but the floor fails Grade 2 for impact sound.

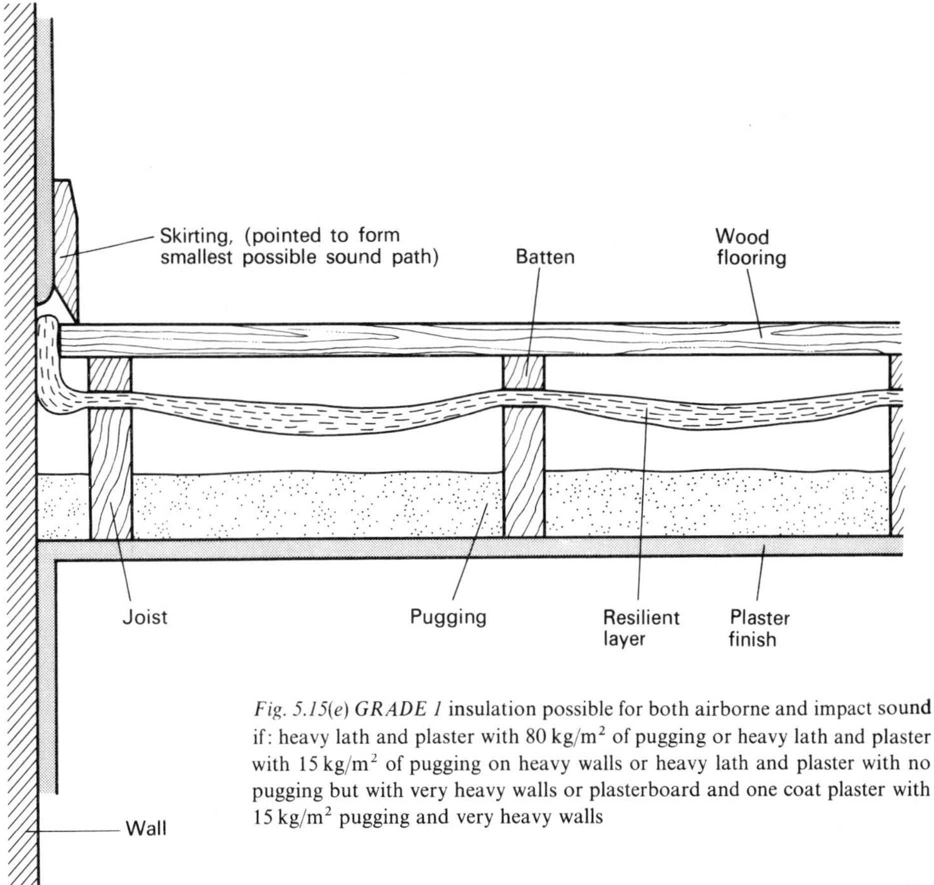


*Fig. 5.15(e) GRADE 1* insulation possible for both airborne and impact sound if: heavy lath and plaster with 80 kg/m$^2$ of pugging or heavy lath and plaster with 15 kg/m$^2$ of pugging on heavy walls or heavy lath and plaster with no pugging but with very heavy walls or plasterboard and one coat plaster with 15 kg/m$^2$ pugging and very heavy walls

## Sound Insulation of Composite Partitions

The sound reduction of a partition measured in dB will depend upon the proportion of sound energy transmitted.

$$\text{Reduction in dB} = 10 \log_{10} \frac{1}{T}$$

Where $T$ is the transmission coefficient.

When calculating the sound insulation of a partition consisting of more than one part (e.g. a 115 mm brick wall with a door) it is first necessary to find the transmission coefficient of each. From this the average transmission coefficient may be calculated using this formula:

$$T_{AV} \times A = T_1 \times A_1 + T_2 \times A_2 + T_3 \times A_3 + \ldots$$

where $T_{AV}$ = average transmission coefficient
$A$ = total area of the partition
$T_1, T_2$ = transmission coefficients of each section
$A_1, A_2$ = areas of each part

### Example 5·4

A partition of total area 10 m$^2$ consists of a brick wall plastered on both sides to a total thickness of 254 mm and contains a door of area 2 m$^2$. The brickwork gives a mean sound reduction of 51 dB and the door 18 dB. Calculate the sound reduction of the complete partition.

*Brickwork*

$$51 = 10 \log_{10} \frac{1}{T_B}$$

where $T_B$ = transmission coefficient of the brick

$$\therefore \quad 5{\cdot}1 = -\log_{10} T_B$$
$$\therefore \quad \log_{10} T_B = \bar{6}{\cdot}9$$
$$\therefore \quad T_B = \text{Antilog}\ \bar{6}{\cdot}9$$
$$= 0{\cdot}000008$$

*Door*

$$18 = 10 \log_{10} \frac{1}{T_D}$$

where $T_D$ = transmission coefficient of the door

$$\therefore \quad 1{\cdot}8 = -\log_{10} T_D$$
$$\therefore \quad \log_{10} T_D = \bar{2}{\cdot}2$$
$$\therefore \quad T_D = 0{\cdot}01585$$

Now:

$$T_{AV} \times 10 = 0{\cdot}000008 \times 8 + 0{\cdot}01585 \times 2$$

$$\therefore \quad T_{AV} = \frac{0{\cdot}000064 + 0{\cdot}031700}{10}$$

$$= 0{\cdot}0031764$$

Actual sound reduction, dB

$$= 10\log_{10}\frac{1}{T_{AV}}$$
$$= -10\log_{10} 0{\cdot}0031764$$
$$= -10 \times \bar{3}{\cdot}5019$$
$$= 30 - 5{\cdot}019$$
$$= 25\,\text{dB}$$

It can be seen that the poor insulation of the door of small area reduces the overall insulation very considerably. If the door had fitted badly the insulation would be even lower.

### Requirements to Achieve Good Sound Insulation

(1) *Mass*
The insulation from a single partition is approximately:

$$R_{AV} = 10 + 14{\cdot}5\log_{10} m$$

where $R_{AV}$ = average sound reduction in dB
$m$ = mass/unit area in $kg/m^2$

The greater the mass, the larger the insulation provided by a partition. A one-brick wall (approximately 230 mm thick) has a mass of 415 $kg/m^2$ and gives an average insulation of about 50 dB. The use of mass up to that of the brick wall is often the most economic method of providing sound insulation. Above an average of 50 dB other methods must be considered. In some cases structural considerations, may prevent the use of mass to provide even moderate sound insulation.

(2) *Completeness*
Example 5·4 showed that the actual insulation of the composite partition was much nearer the value given by the poorer part than that of the brickwork. A wall which might normally have an insulation of 50 dB would have this reduced by means of a hole of only $\frac{1}{100}$ of the total area to about 20 dB. The first consideration must be to try to raise the insulation of the poorest parts, which means that air gaps around doors and windows should be eliminated. Care must be taken where a partition is taken up to the underside of a perforated false ceiling (see Fig. 5.16). This may be a very difficult problem to overcome because of the services above the ceiling. The result can easily be a reduction in the insulation of the partition of 10 dB or more.

(3) *Multiple or Discontinuous Construction*
This is not likely to be economic where the insulation of each leaf of the construction is already high. It would usually then be cheaper to use the same total mass in the form of a single partition. Any improvement in such a case would normally only be in the higher frequency region, where an improvement is less important. Thus a 280 mm cavity brick wall has little advantage over a 230 mm solid brick wall for sound insulation.

Where the panel is light in weight, a double partition may give a definite improvement provided that:

(a) the gap is large, below 50 mm having no advantage
(b) the two panels are of different superficial weight
(c) the gap contains sound absorbent material

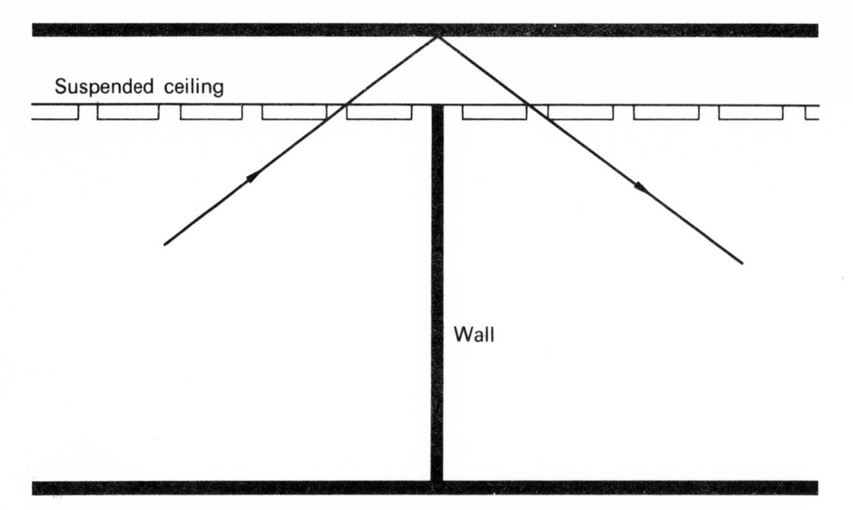


*Fig. 5.16* An example of incomplete construction where a good sound path is present above a perforated false ceiling

(d) there are no air paths through the panels
(e) the panels are not coupled together by the method of construction. It will be fairly clear that if it is necessary to tie the panels together, then to get the greatest impedance mismatch, very light ties should be used with heavy panels, and massive ties with light panels.

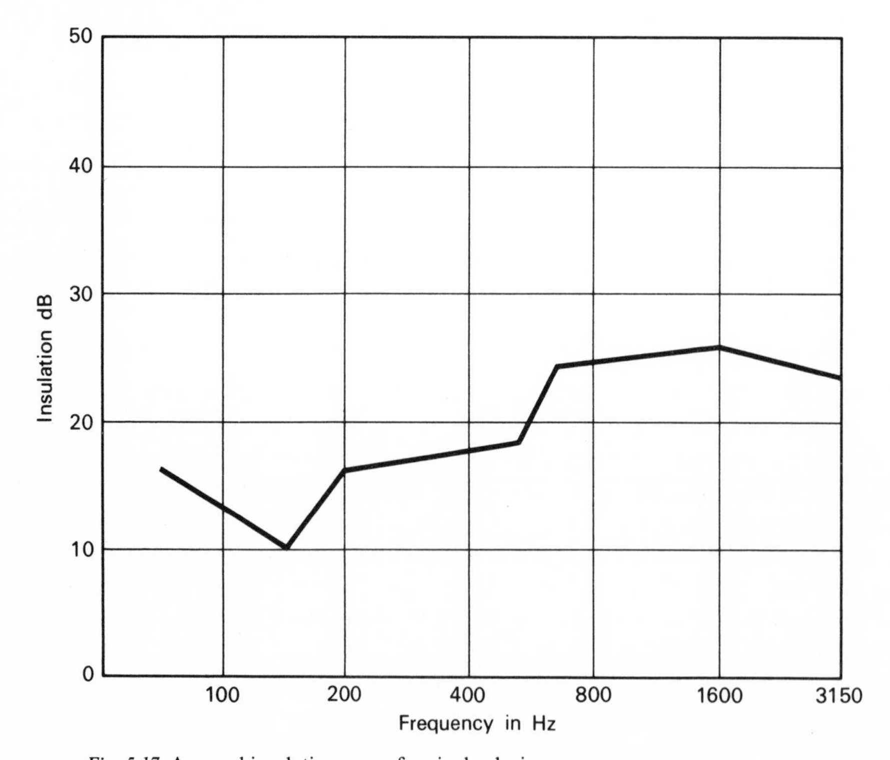


*Fig. 5.17* A sound insulation curve for single glazing

The most common use of double construction is for windows. Typical insulation curves for single and double windows are shown in Figs. 5.17 and 5.18.

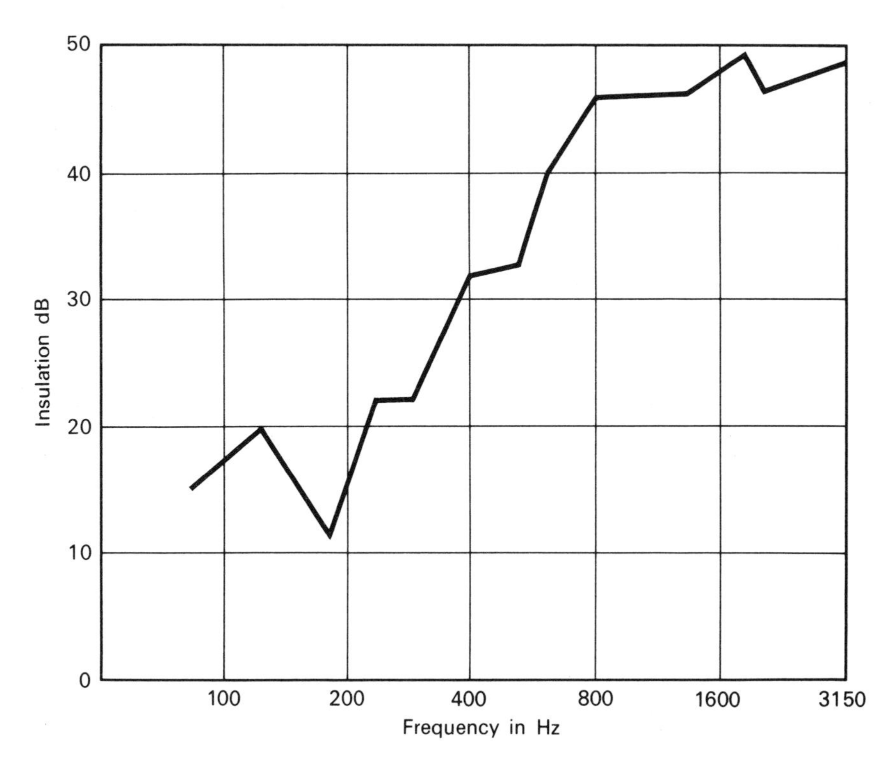


*Fig. 5.18* A sound insulation curve for double glazing

**Table 5·3**

*Sound Reduction in dB for Windows*

| Sound reduction index (dB) at frequency Hz | | | | | | | |
|---|---|---|---|---|---|---|---|
| Construction | 100 | 200 | 400 | 800 | 1600 | 3150 | Mean |
| 1. 2·5 mm glass about 500 × 300 mm in metal frames, closed openable sections | 23 | 12 | 18 | 22 | 29 | 27 | 21 |
| 2. Same as 1. but sealed | 18 | 12 | 20 | 27 | 30 | 28 | 23 |
| 3. Same as 1. but wood frames | 14 | 17 | 22 | 26 | 29 | 30 | 23 |
| 4. Same as 1. but 6 mm glass in wood frames | 16 | 20 | 24 | 29 | 29 | 36 | 26 |

**Table 5·3**

*Sound Reduction in dB for Windows*

| Sound reduction index (dB) at frequency Hz | | | | | | | |
|---|---|---|---|---|---|---|---|
| Construction | 100 | 200 | 400 | 800 | 1600 | 3150 | Mean |
| 5. 25 mm glass 650 × 540 mm in wood frames | 28 | 25 | 30 | 33 | 41 | 47 | 34 |
| 6. Same as 1. but double wood frames in brickwork, 50 mm cavity between glazing | 19 | 19 | 27 | 37 | 47 | 52 | 33 |
| 7. Same as 6. but 100 mm cavity | 22 | 20 | 35 | 43 | 51 | 53 | 37 |
| 8. Same as 6. but 180 mm cavity | 28 | 25 | 36 | 43 | 50 | 53 | 39 |
| 9. Same as 8. but with acoustic tiles on reveals | 25 | 34 | 41 | 45 | 53 | 57 | 42 |

(4) *Apparent Insulation by the Use of Absorbents*
The necessity for a correction to be applied to the measured value of sound insulation shows that the sound pressure level in a room is to some extent dependent upon the reverberation time of that room. The improvement that can be achieved by the use of absorbent materials in dB $= 10 \log_{10} (A+a)/A$

where $A$ = number of $m^2$ of absorption originally
$a$ = number of $m^2$ of absorption added

The improvement in a highly reverberant room can be considerable, but in a room which already has a large amount of absorption, little improvement is likely to be possible.

**Example 5·5**

Using the Sabine formula calculate the area of absorption that needs to be added to a room of volume 100 $m^3$ whose reverberation time was originally 2·00 s to give:

(a) 3 dB reduction in the sound pressure level
(b) 6 dB reduction in the sound pressure level
(c) 10 dB reduction in the sound pressure level

Amount of absorption originally:

$$2{\cdot}00 = \frac{0{\cdot}16 \times 100}{A}$$

$$\therefore \quad A = 8 \text{ m}^2$$

(a) $$3 = 10\log_{10}\frac{8+a_1}{8}$$

$$\therefore\quad 0{\cdot}3 = \log_{10}\frac{8+a_1}{8}$$

$$\therefore\quad \frac{8+a_1}{8} = 2$$

$$a_1 = 8\ \text{m}^2$$

(b) $$6 = 10\log_{10}\frac{8+a_2}{8}$$

$$\therefore\quad \frac{8+a_2}{8} = 4$$

$$a_2 = 24\ \text{m}^2$$

(c) $$10 = 10\log_{10}\frac{8+a_3}{8}$$

$$1 = \log_{10}\frac{8+a_3}{8}$$

$$\therefore\quad \frac{8+a_3}{8} = 10$$

$$8+a_3 = 80$$

$$a_3 = 74\ \text{m}^2$$

It can be seen that a tenfold increase is needed for 10 dB reduction. For 3 dB the amount of absorption must be doubled.

Absorbent materials can be very useful near a source of sound because the amount needed can be fairly small. They can often be a way of localizing noise in factories. This is not a means of insulation and the improvement is not nearly as good but nevertheless can be useful. In a large factory it can improve safety because a machine operator can hear his own machine more clearly than others (see Fig. 5.19).

The use of absorbents in ducts and corridors can prevent noise reaching quiet rooms.

### Attenuation in Ducts

Noise from ventilator fans or from other rooms can easily be transmitted by means of ducts. By lining the duct with absorbent an attenuation per metre, $R_1$, given approximately by:

$$R_1 = \left(\frac{P}{S}\right)\alpha^{1{\cdot}4} \text{ is obtained}$$

where $P$ = perimeter of the duct in metres
$S$ = cross-sectional area in $\text{m}^2$
$\alpha$ = absorption coefficient

This equation is accurate enough for most practical purposes provided that the area of the duct does not exceed about $0{\cdot}3\ \text{m}^2$ for ducts which are more nearly square than 2:1.

Increased attenuation can be obtained by means of duct splitters running along all or part of the duct effectively making it a number of small parallel ducts. The chief

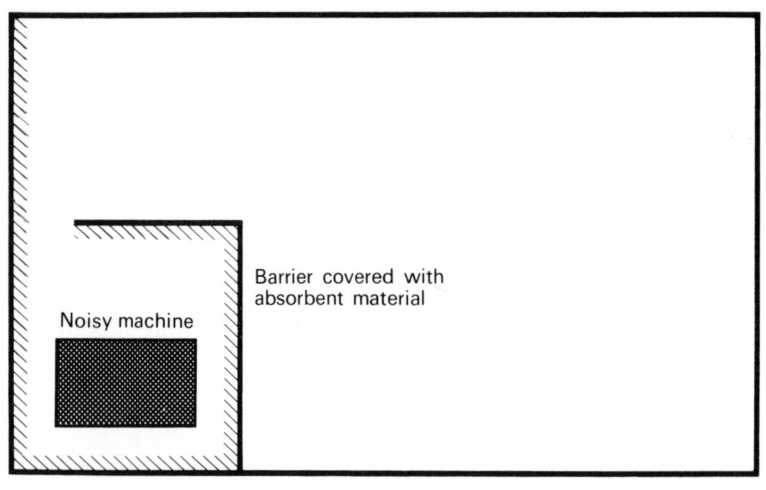


*Fig. 5.19* Use of absorbent covered screen can help localize noise. Absorbent material is necessary to prevent sound being reflected back to the operator of the machine, raising the level there even more

disadvantage of this is that it cuts the total cross-sectional area with the result that the air speed must be increased.

### Grille Attenuation

Reduction of sound at the opening is dependent upon the area of the grille and the total sound absorption within the room.

$$R_2 = 10 \log_{10} \frac{A}{S}$$

where $A$ = room absorption in $m^2$ units
$S$ = area of the grille in $m^2$

As the absorption within a room may vary with frequency it may be necessary to make a number of calculations for grille attenuation.

### Example 5·6

A fan for a hot air heating system is found to produce the following noise:

| Frequency Hz | 75–150 | 150–300 | 300–600 | 600–1200 |
|---|---|---|---|---|
| Level, dB | 90 | 85 | 80 | 75 |

The duct is 0·2 m × 0·4 m internally with a lining of absorbent whose coefficient is:

| Frequency Hz | 75–150 | 150–300 | 300–600 | 600–1200 |
|---|---|---|---|---|
| Absorption coefficient, $\alpha$ | 0·1 | 0·2 | 0·5 | 0·85 |

Calculate the level in a room of volume 500 $m^3$ with a reverberation time of 2·00 s if the duct is 4 m long and straight with a grille the same area as the duct.

| Frequency | Fan level dB | $\alpha$ | $\alpha^{1\cdot 4}$ | $4\frac{P}{S}\alpha^{1\cdot 4}$ | $10\log_{10}\frac{A}{S}$ | Room level dB |
|---|---|---|---|---|---|---|
| 75–150 | 90 | 0·1 | 0·05 | 3 | 27 | 63 |
| 150–300 | 85 | 0·2 | 0·10 | 6 | 27 | 52 |
| 300–600 | 80 | 0·5 | 0·38 | 22·8 | 27 | 30·2 |
| 600–1200 | 75 | 0·85 | 0·80 | 48 | 27 | 0 |

Sabine's formula: $t = \dfrac{0\cdot 16\, V}{A}$

$$A = \frac{0\cdot 16 \times 500}{2\cdot 00}$$

$$= 40\ \mathrm{m}^2 \qquad \left(\text{Hence } 10\log_{10}\frac{A}{S}\right)$$

## Attenuation Due to Bends

The amount of attenuation at a bend depends upon frequency, size of duct and whether it is lined or unlined. An approximate relation is given in Table 5·4.

**Table 5·4**

*Attenuation at Right-Angled Bend for Lined Square Ducts*

| Octave band centre frequency Hz | 0·1 $m^2$ Area (square) | 1 $m^2$ Area (square) |
|---|---|---|
| 63 | 0 | 1 |
| 125 | 1 | 3 |
| 250 | 3 | 8 |
| 500 | 6 | 16 |
| 1000 | 8 | 17 |
| 2000 | 16 | 18 |
| 4000 | 17 | 18 |
| 8000 | 18 | 18 |

## Sound Power Levels of Ventilation Fans

Information about the sound power level in each octave band will normally be supplied by the manufacturer. Alternatively, measurements may be made using a sound level meter and microphone equipped with a windshield. In the absence of the necessary data it is possible to estimate the values approximately from the formula:

$$L_w = 61 + 10\log_{10} W$$

where $L_w$ = sound power level in dB re 1 picowatt
$W$ = power of the motor in watts

This formula assumes that the motor is run at full power. If a fan is run at half its rated power then a reduction of 6 dB will result. If a larger motor is used of twice the power the formula shows that an increase of 3 dB is expected.

The spectra of sound depends upon the type of fan being used. It can be seen from Fig. 5.20 that centrifugal fans produce most of their noise in the low frequencies whereas axial flow fans have a far more even distribution.

The sound pressure level in a room may be calculated from:

$$\text{S.P.L.} = L_w - 10\log_{10}\left(\frac{A}{4}\right)$$

where $A$ = area of room absorbent in $m^2$

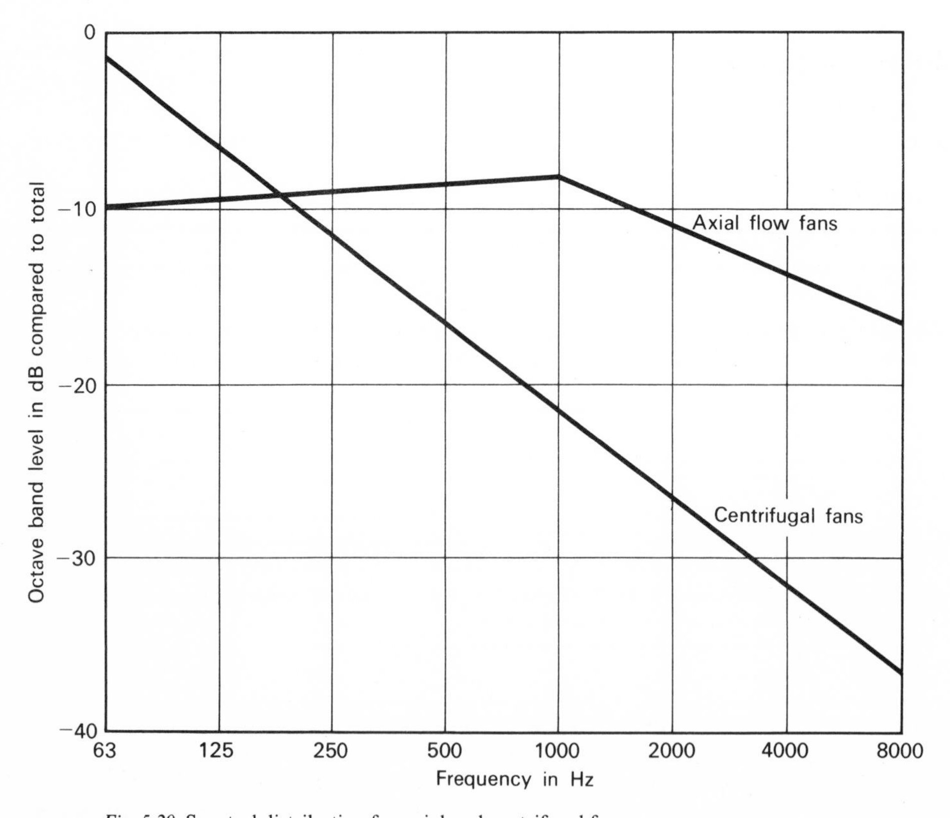


*Fig. 5.20* Spectral distribution for axial and centrifugal fans

When the sound power level is known, the distribution through the duct system can be calculated taking into account that where ducts split the energy will be divided. If the maximum permissible level in a room is known the attenuation needed within the system can be found.

## Questions

(1) The noise level from a sewing machine was found to be 96 dB (linear) at a distance of 1 metre. Find the sound power of the machine in picowatts. (Assume it is in a free field.)

(2) A lighthouse has a foghorn with a sound power output of 100 watts. If it radiates over a quarter of a sphere, at what distance would the level be 100 dB? Why would a low frequency horn be better than one of higher frequency?

(3) Calculate the sound output of a man whose loud speech produces a sound pressure level of 80 dB(c) at a distance of 1 metre.

(4) (a) Explain what is meant by each of the terms 'decibel' and 'sound reduction index'.
(b) An external wall, of area 4 m by 2·5 m in a house facing a motorway is required to have a sound reduction of 50 dB. The construction consists of a 280 mm cavity wall containing a double glazed (and sealed) window. The sound reduction indices are 55 dB for the 280 mm cavity wall and 44 dB for the sealed double glazed window with a 150 mm cavity. Calculate to the nearest 0·1 $m^2$ the maximum size of window to achieve the required insulation.

(5) (a) Explain how sound is transmitted from one room to another in a building and show how it can be reduced.
(b) A partition across a room 6 m by 5 m includes a timber doorway of area 2 $m^2$. Determine by how much the partition will be a better insulator of sound if constructed in 115 mm brickwork instead of 100 mm building blocks, each plastered both sides, at the particular frequency where the sound reduction indices of brick, building block and timber are respectively 48 dB, 34 dB and 26 dB.

(6) (a) Name four principles to be observed when considering sound insulation and say whether they are effective against air borne or structure borne sound.
(b) Show how to apply these principles in dividing a room into a consultant's office and waiting room.
(c) Calculate the average sound reduction factor of a partition made of 24 $m^2$ of 115 mm brickwork and 6 $m^2$ of plate glass where the sound reduction factors of brickwork and glass at a certain frequency are 45 dB and 30 dB respectively.

(7) A room of volume 200 $m^3$ is supplied with air from a centrifugal fan of 400 watts. Find the amount of attenuation needed in each octave band. The reverberation times of the room and the maximum acceptable levels at the different frequencies are:

| Octave band centre frequency Hz | 63 | 125 | 250 | 500 | 1000 | 2000 | 4000 |
|---|---|---|---|---|---|---|---|
| Reverberation time in s | 5 | 3 | 1·5 | 0·6 | 0·5 | 0·5 | 0·6 |
| Maximum acceptable levels dB | 66 | 59 | 52 | 46 | 42 | 40 | 38 |

(8) Calculate the attenuation provided by a 0·3 m square duct 3 m long and lined with material having the following absorption coefficients:

| Octave band centre frequency Hz | 63 | 125 | 250 | 500 | 1000 | 2000 | 4000 |
|---|---|---|---|---|---|---|---|
| $\alpha$ | 0·10 | 0·37 | 0·47 | 0·70 | 0·80 | 0·85 | 0·80 |

(9) Calculate the sound reduction of a partition of total area 20 $m^2$ consisting of 15 $m^2$ of brickwork, 3 $m^2$ of windows and 2 $m^2$ of door. The sound reduction indices are 50 dB, 20 dB and 26 dB for the brickwork, windows and door respectively.

(10) A workshop of volume 1000 $m^3$ has a reverberation time of 3 seconds. The sound pressure level with all the machines in use is 105 dB at a certain frequency. If the reverberation time is reduced to 0·75 seconds what would the sound pressure level be?

(11) Find the reduction in sound pressure level of a 165 Hz frequency by a barrier of height 4 m alongside a motorway if the noise source is 500 mm above the base of the barrier and 5 m away from it. The listener is 1 m below the base and 10 m away from it.

(12) If an average sound reduction of 20 dB is required by a barrier, what is the approximate minimum mass per unit area required?

Chapter 6

# Criteria

The optimum noise environment must depend upon both subjective and economic limitations. To find people's requirements necessitates a consideration of three main factors: the noise level, the individual and his activity. It is impossible to make allowances for individual differences and criteria must be stringent enough to ensure that nearly everyone is satisfied. There may be multiple activities within one part of a building, such as a hospital ward. Noise requirements are activity dependent, so a noisy data processing office is not going to have the same criteria as an executive office, for example.

There are three main bases for the determination of criteria: hearing damage, speech interference and annoyance. The first two are fairly easy to determine, whereas annoyance is far less clearcut.

## (1) Deafness Risk Criteria

A sound pressure level of about 150 dB would result in instant hearing damage. People must not be exposed to noises of this magnitude, and in fact it is unwise to exceed 140 dB even for noises of under one minute duration. Fortunately noise of this level very rarely occurs in buildings. Of far greater importance are the long term effects of noise, particularly in industry. Table 2·3 gave a suggested limit for occupational noise 8 hours a day, 5 days a week for 40 years. An increase in these levels is possible if the period of exposure each day is reduced, whereas a decrease would be necessary if the sound was of very narrow bandwidth, as may sometimes occur in industry.

Where it is necessary to exceed the criteria, hearing conservation measures need to be taken. These include the use of ear plugs or ear muffs, shorter exposure and regular audiometric tests (see Fig. 6.1).

## (2) Speech Interference Criteria

Necessity for speech communication will set a more severe limit on the maximum levels of ambient noise than hearing damage. Clarity of speech in the presence of masking noises is dependent not only on their magnitude, but also upon the loudness of the speaker's voice and the hearing acuity of the listener. Little allowance can be made for the variations in the last two and an average is taken. The noise spectra will also determine the interfering effect. The usual method involves taking the arithmetic mean of the sound pressure levels in the three octave bands 600–1200 Hz, 1200–2400 Hz and 2400–4800 Hz. This average is known as the SPEECH INTERFERENCE LEVEL. Distance between speaker and listener is important. Table 6·1 gives permissible speech interference levels (SIL) for different distances. These are for a normal voice and if the level is raised an increase in the speech interference level is permissible.

**Table 6·1**

*Speech Interference Levels*

| Distance from speaker to hearer m | Normal voice dB |
|---|---|
| 0·1 | 73 |
| 0·2 | 69 |
| 0·3 | 65 |
| 0·4 | 63 |
| 0·5 | 61 |
| 0·6 | 59 |
| 0·8 | 56 |
| 1·0 | 54 |
| 1·5 | 51 |
| 2·0 | 48 |
| 3·0 | 45 |
| 4·0 | 42 |

Raised voice: add 6 dB to each

Very loud voice: add 12 dB to each

Shouting: add 18 dB to each

In the case of a female voice all levels should be reduced by 5 dB

It is possible for the speech interference level to be below the levels shown in Table 6·1 but for the noise to interfere because of the presence of excessive low frequency components. To overcome this difficulty the LOUDNESS LEVEL in phons is also calculated. This is defined as the sound pressure level of a 1000 Hz note that sounds equal in loudness to the noise concerned. If the difference in the loudness level and the speech interference level is less than 22 there will be general satisfaction with the noise environment provided the speech interference level is below the appropriate maxima. However, if the loudness level minus the speech interference level is greater than 30 there will be general dissatisfaction even if the speech interference level is low enough. This is due to the presence of low frequency noise.

Noise criteria (NC) curves are used to specify the maximum speech interference level with the added fact that the difference between loudness level and the speech interference level is equal to 22 (see Fig. 6.2). If lower standards for the aural environment are acceptable then the alternative noise criteria (NCA) curves may be used for specification. These allow the presence of more low frequency noise and are calculated on the basis of the loudness level minus the speech interference level being equal to 30 (see Fig. 6.3).

**Example 6·1**

Calculate the sound insulation needed to achieve the
(a) NC 25 curve
(b) NCA 25 curve
in a classroom if there is an external noise of:

| Octave band centre frequency Hz | 63 | 125 | 250 | 500 | 1000 | 2000 | 4000 | 8000 |
|---|---|---|---|---|---|---|---|---|
| Level dB | 75 | 71 | 70 | 69 | 65 | 62 | 61 | 60 |

(a) To achieve NC 25 level

| Frequency Hz | Level dB | NC 25 (from Fig. 6.2) | Insulation dB |
|---|---|---|---|
| 63 | 75 | 57 | 18 |
| 125 | 71 | 47 | 24 |
| 250 | 70 | 39 | 31 |
| 500 | 69 | 32 | 37 |
| 1000 | 65 | 28 | 37 |
| 2000 | 62 | 25 | 37 |
| 4000 | 61 | 22 | 39 |
| 8000 | 60 | 21 | 39 |

(b) To achieve NCA 25 level

| Frequency Hz | Level dB | NCA 25 (from Fig. 6.3) | Insulation dB |
|---|---|---|---|
| 63 | 75 | 66 | 9 |
| 125 | 71 | 53 | 18 |
| 250 | 70 | 42 | 28 |
| 500 | 69 | 34 | 35 |
| 1000 | 65 | 28 | 37 |
| 2000 | 62 | 25 | 37 |
| 4000 | 61 | 22 | 39 |
| 8000 | 60 | 21 | 39 |

Each of these standards could be achieved by means of a 115 mm brick wall.

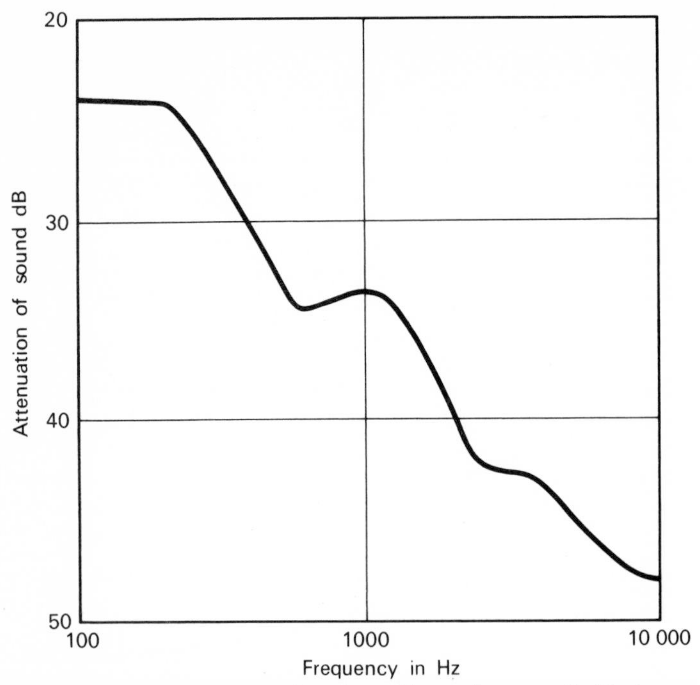


*Fig. 6.1* Typical attentuation of sound by wearing ear muffs. Maximum attenuation is in the high frequency region

**Example 6·2**

The recommended maximum sound level for the sleeping area of a house is the NC 35 curve. The external bedroom wall of a house is 4 m by 2·5 m in area, facing a motorway at a distance of 30 m from the line of traffic. An octave band analysis at 7·5 m from the line of traffic gave the following results:

| Frequency Hz | 20–75 | 75–150 | 150–300 | 300–600 | 600–1200 | 1200–2400 | 2400–4800 | 4800–10 000 |
|---|---|---|---|---|---|---|---|---|
| | 99 | 95 | 93 | 92 | 93 | 90 | 88 | 85 |

The construction consists of a 280 mm cavity wall containing a double glazed (and sealed) window. The sound reduction indices are 55 dB for the 280 mm cavity wall and 44 dB for the sealed double window both at 1000 Hz. Assume for the purpose of this question that the insulation of both the window and brickwork improves by 5 dB for each doubling of the frequency. Calculate, for 1000 Hz, the maximum area of window to the nearest 0·1 $m^2$. State what the insulation is expected to be at the other seven octave values and whether they are likely to be adequate.

The level of the sound will drop 3 dB for each doubling of the distance for noise from a line source. In this example the distance has been doubled twice so that $2 \times 3 = 6$ dB must be subtracted to find the level immediately outside the window.

| Frequency Hz | 20–75 | 75–150 | 150–300 | 300–600 | 600–1200 | 1200–2400 | 2400–4800 | 4800–10 000 |
|---|---|---|---|---|---|---|---|---|
| Motorway dB | 99 | 95 | 93 | 92 | 93 | 90 | 88 | 85 |
| Outside window dB | 93 | 89 | 87 | 86 | 87 | 84 | 82 | 79 |

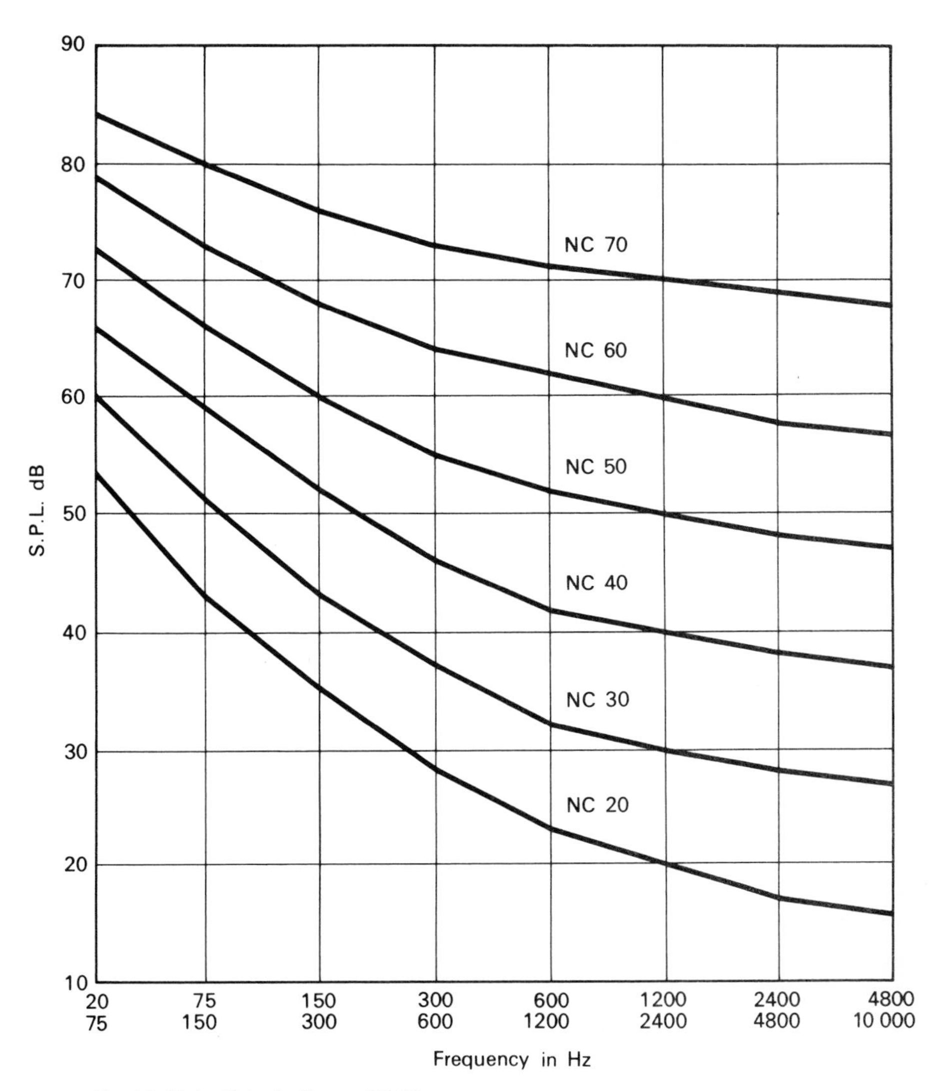


*Fig. 6.2* Noise Criteria Curves (N.C.)

To find the necessary insulation the maximum recommended level given by the NC 35 curve must be subtracted from the values outside the window. Hence:

| Frequency Hz | 20–75 | 75–150 | 150–300 | 300–600 | 600–1200 | 1200–2400 | 2400–4800 | 4800–10 000 |
|---|---|---|---|---|---|---|---|---|
| Outside window dB | 93 | 89 | 87 | 86 | 87 | 84 | 82 | 79 |
| NC 35 | 63 | 55 | 47 | 42 | 37 | 35 | 33 | 32 |
| Insulation needed dB | 30 | 34 | 40 | 44 | 50 | 49 | 49 | 47 |

Thus it can be seen that the necessary insulation at 1000 Hz must be 50 dB.

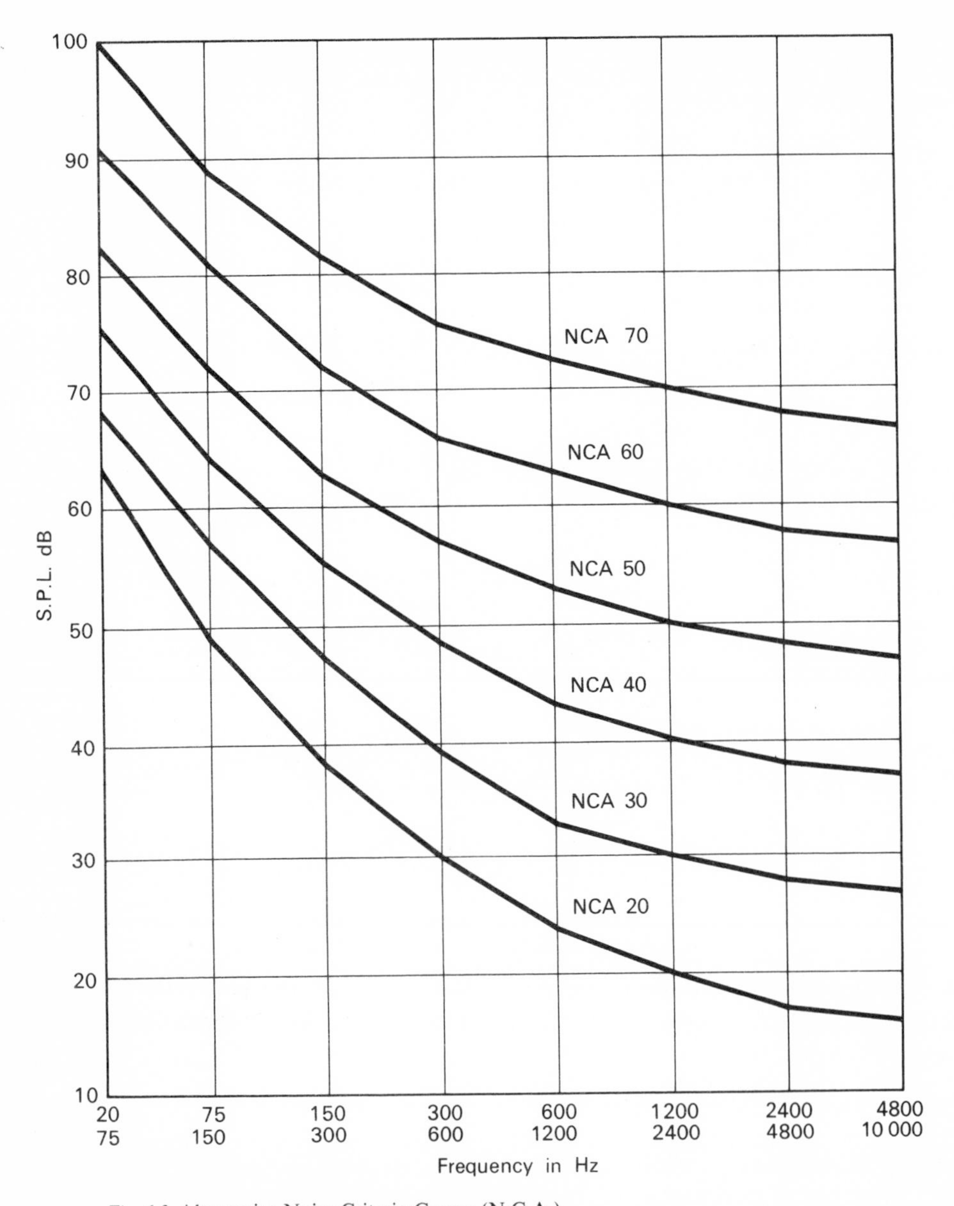


*Fig. 6.3* Alternative Noise Criteria Curves (N.C.A.)

Now the reduction in dB = $10 \log_{10} (1/T)$ where $T$ is the transmission coefficient

$$\therefore \quad 50 \text{ dB} = 10 \log_{10} \frac{1}{T_{AV}}$$

($T_{AV}$ = average transmission coefficient for the window and brickwork combined)

$$\therefore \quad 5 = \log_{10} \frac{1}{T_{AV}}$$

$$= -\log_{10} T_{AV}$$

$$\therefore \quad \bar{5}\cdot 0000 = \log_{10} T_{AV}$$

$$T_{AV} = 0\cdot 00001$$

For the 280 mm cavity wall:

$$55\ \text{dB} = 10 \log_{10} \frac{1}{T_B}$$

($T_B$ = transmission coefficient for the brickwork)

$$\therefore \quad 5\cdot 5 = -\log_{10} T_B$$

$$\bar{6}\cdot 5 = \log_{10} T_B$$

$$T_B = 0.000003162$$

Similarly for the window:

$$44 = 10 \log_{10} \frac{1}{T_W}$$

($T_W$ = transmission coefficient for the window)

$$\therefore \quad 4\cdot 4 = -\log_{10} T_W$$

$$\bar{5}\cdot 6 = \log_{10} T_W$$

$$T_W = 0\cdot 00003981$$

Now:

$$(\text{Total area}) \times T_{AV} = (\text{area of brick}) \times T_B + (\text{area of window}) \times T_W$$

Let the window area $= a\ \text{m}^2$

Total area $= 10\ \text{m}^2$

Area of brick $= (10 - a)\ \text{m}^2$

$$\therefore \quad 10 \times 1\cdot 0 \times 10^{-5} = (10 - a) \times 3\cdot 162 \times 10^{-6} + a \times 3\cdot 981 \times 10^{-5}$$

$$1\cdot 0 \times 10^{-4} = 3\cdot 162 \times 10^{-5} - 3\cdot 162 \times 10^{-6} a + 3\cdot 981 \times 10^{-5}$$

$$\therefore \quad 6\cdot 838 \times 10^{-5} = 3\cdot 665 \times 10^{-5} a$$

$$a = \frac{6838}{3665}$$

$$= 1\cdot 9\ \text{m}^2 \text{ to the nearest } 0\cdot 1 \text{ square metre}$$

A 5 dB improvement in insulation for each doubling of frequency also means a reduction of 5 dB for each halving of the frequency below 1000 Hz. Thus the expected insulation would be:

| Hz | Octave band | Insulation dB |
|---|---|---|
| 2000 | 1200–2400 | 55 |
| 4000 | 2400–4800 | 60 |
| 8000 | 4800–10 000 | 65 |

| | | |
|---|---|---|
| 500 | 300–600 | 45 |
| 250 | 150–300 | 40 |
| 125 | 75–150 | 35 |
| 63 | 20–75 | 30 |

Comparing these values with those from the NC curve the insulation should be adequate for all frequencies.

### Criteria for Offices

Speech interference levels were developed to give office criteria. The use of an NC or NCA curve is therefore convenient. Selection of a particular NC or NCA curve is dependent upon the type and use of the office, combined with the economic requirements. NCA curves should not be used unless cost considerations prevent the use of NC curves. Levels of background noise from NC 20 for the best executive offices to NC 55 in offices containing typewriters or other machinery are suitable.

It seems that for certain types of offices higher levels would be acceptable where speech is not important. In data processing offices levels of 79 dBA for machine operators, 72 dBA for punch operators and 64 dBA for other clerical staff appear to be satisfactory. The main difference in the latter cases is that the noise is under the control of the operators and is not an external sound.

### Audiometric Rooms

In order that audiometric rooms can measure accurately the amount of hearing loss, if any, that a person suffers, it is important that there is no masking noise. A criteria must be set at such a level that it is possible to measure the threshold of hearing for people with normal hearing (zero hearing loss). Fig. 6.4 shows the maximum permissible levels measured in octave bands for this to be achieved. If the levels are measured in half or third octave bands the criteria must be set correspondingly lower.

The criteria for third octave bandwidth measurements at each frequency are $10 \log_{10} 3 = 10 \times 0{\cdot}4771 \simeq 5$ dB lower, and for half octaves, $10 \log_{10} 2 = 3$ dB lower.

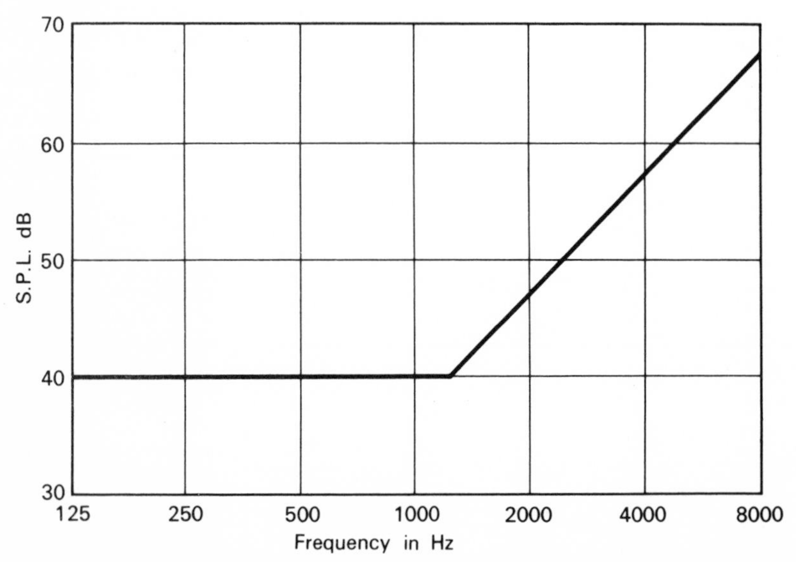


*Fig. 6.4* Maximum sound pressure levels for no masking above zero hearing loss

If narrow band sound is present then the criteria given can be satisfied, but masking may be caused at the audiometric measuring frequencies. It is necessary, therefore, to check that there is no narrow band sound at frequencies near the test frequency. For further information on this the reader is referred to bibliographical reference 1.

Where the audiometric room is only to be used for testing people with some hearing loss, higher levels of background noise are permissible. This means that for testing people with a 10 dB hearing loss, the criteria can be relaxed by approximately 10 dB at each frequency, and similarly 20 dB higher for a 20 dB hearing loss. It also applies, however, that if the room is to be used to measure the hearing acuity of people with better than average hearing, a more severe criteria would be needed.

### Hospitals

There are no standards at present relating to noise or noise control in hospitals. There are reports from various countries suggesting criteria. The NC 30 curve would appear to be in general agreement with these recommended maximum criteria for hospital wards (see Table 6·2).

Hospitals probably represent one of the most difficult environments for which to obtain satisfactory criteria. Of necessity they must have a considerable quantity of mechanical aids. At the same time some people are sleeping whilst others are working. Some internal ward noise is inevitable. If internal noise is reduced to a too low level by means of insulation the masking effect it provides is lost and internal noises are more noticeable. In large wards there is an additional problem, in very quiet conditions, of patients hearing doctors discussing their diagnosis of other patients. Taking this need for masking into account, daytime levels from all sources within the range 45–50 dBA have been provisionally recommended (Bibliographical reference 8).

Night time noise may represent a severe problem at far lower levels than those recommended. In the absence of external masking noises structure borne noises can become far more serious. Care in placing machinery in the planning stage is needed to prevent this becoming disturbing to patients. Machinery should be selected with due regard to its noise level as well as to cost (Bibliographical references 9 and 10).

**Table 6·2**

*Suggested Criteria for Hospitals*

| | |
|---|---|
| Bedrooms | NC 30 |
| Day rooms | NC 35 |
| Treatment rooms | NC 35 |
| Bathrooms | NC 35 |
| Toilets | NC 35 |
| Kitchens | NC 40 |

## Domestic Dwellings

Ideally the level in the sleeping area of homes should not exceed the NC 25–35 curves. One problem in dwelling houses will be the noise from neighbours. A correctly specified minimum wall insulation should make certain that, with the exception of particularly noisy neighbours, the desired reduction is achieved. Grade curves for airborne and impact sound insulation were given in Figs. 4.8 and 4.9 and Figs. 5.13 and 5.14.

To achieve a greater insulation would be uneconomic and probably undesirable as housewives like to be aware of a small amount of neighbourhood noise during the daytime. In any case noise produced by equipment within an individual dwelling will provide some masking effect.

## Classrooms and Lecture Rooms

The problem is primarily one of speech interference. Ideally a level no greater than the NC 25 curve should be achieved for continuous noise. It appears that higher levels can be satisfactory, provided they do not exceed the NC 35 curve. Where the noise is intermittent and infrequent, such as is the case with railway and aircraft noise, lower standards may be tolerable. When the noise occurs every few minutes or more frequently the interruption can have serious effects on the rate of learning.

A recent development which should cut down intermittent noise involves windows which are automatically closed when the external noise level exceeds a predetermined amount. This system is basically a microphone located on the roof of the school controlling a hydraulic servomechanism to close all the double glazed windows. The advantage of this method is that ventilation problems are overcome. The alternative of fixed double glazed windows requires a system of forced ventilation which will itself produce noise exceeding the NC 25 level.

Music room requirements are similar to those of classrooms, but present a problem if situated close to other teaching rooms. The level of sound produced even by a small school orchestra can be fairly high. If it is necessary to situate a music practice room near to a teaching room, then increased insulation is advisable, equivalent to a 230 mm brick wall.

## Hotels

The requirements for hotel bedrooms are similar to those for domestic dwellings with a recommended level of around NC 25. This is quite impossible economically in certain places, and slightly higher levels up to about NC 35 or 40 should be acceptable.

## Broadcasting and Recording Studios

Different research workers have suggested various criteria including those lying below the threshold of hearing at low frequencies, and below the level of self generated noise of a high quality condenser microphone at higher frequencies. (This exceeds the threshold of hearing at high frequencies.) The levels given in Table 6·3 would be more reasonable.

**Table 6·3**

*Maximum Permissible Background Noise Levels in Studios*

| Octave band Hz | Sound only Light entertainment dB | Drama dB | Other sound All television dB |
|---|---|---|---|
| 37–75 | 55 | 45 | 50 |
| 75–150 | 45 | 36 | 40 |
| 150–300 | 38 | 27 | 32 |
| 300–600 | 32 | 23 | 27 |
| 600–1200 | 27 | 18 | 22 |
| 1200–2400 | 23 | 14 | 17 |
| 2400–8000 | 20 | 10 | 14 |

### Concert Halls

The best concert halls should have a background noise no higher than the threshold of hearing for continuous noise. This level is shown approximately in Table 6·4. In many cases this standard is too costly and a lower specification must be accepted but not worse than the NC 20 curve. Opera houses and large theatres with more than 500 seats should achieve NC 20.

**Table 6·4**

*Maximum Background Noise Levels for Concert Halls*

| Octave band centre frequency Hz | 63 | 125 | 250 | 500 | 1000 | 2000 | 4000 | 8000 |
|---|---|---|---|---|---|---|---|---|
| Sound pressure level dB | 53 | 38 | 28 | 18 | 12 | 11 | 10 | 22 |

### Churches

Traditional massive construction in Britain has provided most churches with very high standards of insulation. Increase in road traffic combined with lighter construction is making it more difficult to achieve the desirable noise level in modern buildings. Lighter construction with larger windows has increased the ventilation problem, necessitating open windows and further reducing the sound insulation. The NC 25 curve should represent a suitable level if there is no means of amplification. On a busy main road site this will be difficult to achieve.

## Industrial Noise Affecting Residential Areas

The main problem with most industrial noise has already been discussed in relation to hearing loss. However, there is an additional problem where noise from an industrial process will be heard by people in their homes. A limit must be set to the disturbing effect of such a noise, taking due account of its character. BS 4142 sets a basic criteria of 50 dBA to which adjustments are made according to circumstances.

Briefly the method of obtaining the criteria is:

1. Basic criteria 50 dB
2. Add:
   (a) 0 dB for new factories;
   (b) 0 dB for existing factories to which changes of structure or process are being made;
   (c) 5 dB for existing factories out of character with the neighborhood;
   (d) 10 dB for long existent factories in keeping with the neighbourhood.
3. To allow for type of district:
   (a) subtract 5 dB for rural area;
   Add:
   (b) 0 dB for suburban area;
   (c) 5 dB for urban area;
   (d) 10 dB for mainly residential urban area with some light industry or main roads;
   (e) 15 dB for evenly mixed residential and industrial area;
   (f) 20 dB for a predominantly industrial region with few dwellings.
4. To allow for times of day that the factory is in use:
   (a) Weekdays 8·00–18·00 hours only, add 5 dBA;
   (b) To include night time 22·00–7·00 hours, subtract 5 dBA;
   (c) If in use between the periods given in (a) and (b), but NOT in (b), then no change is needed.
5. To allow for seasonal correction add 5 dBA if only in use during the winter — when windows are more normally closed.

Together with this calculation of the criteria is the need for measurement or assessment of the noise level with due allowance for its character. This is done by:

1. Finding the noise level in dBA;
2. Applying a tonal character adjustment of +5 dB where the noise has a definite continuous note such as a whine or hiss;
3. Adding 5 dB if the sound is impulsive, e.g. bangs;
4. Applying a correction to allow for the intermittent nature of the sound. This is done by finding the percentage of the time the noise is produced and applying a correction from Fig. 6.5 or Fig. 6.6, depending whether it is day or night.

The corrected noise level may then be compared with the criteria obtained earlier. If this corrected noise level exceeds the criteria by 10 dBA or more, complaints may be expected. When the level is 10 dBA or more below the criteria no complaints should arise.

When complaints are received about industrial noise and are being investigated it is useful to make both objective measurements as described and also a subjective

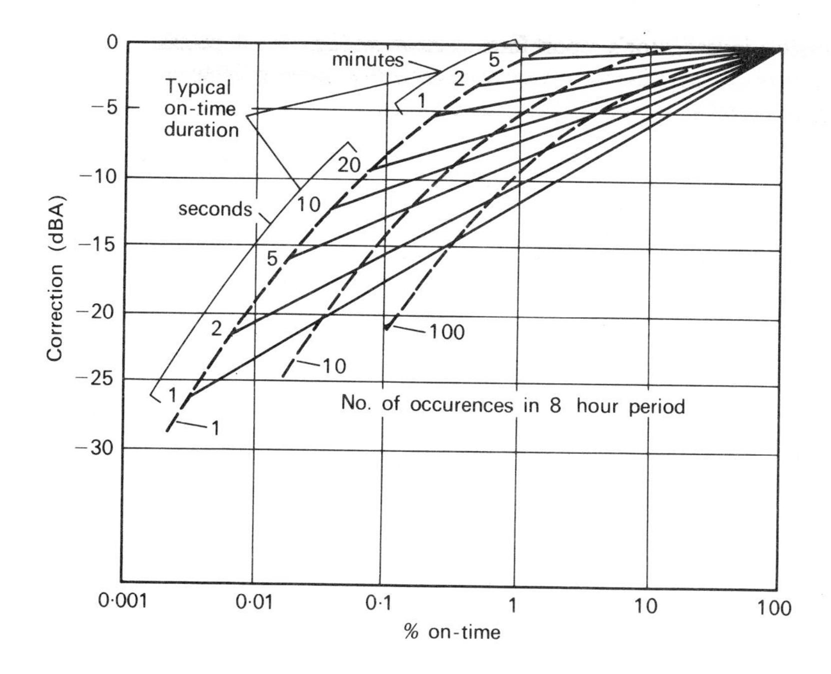


*Fig. 6.5* Intermittency duration correction for night time

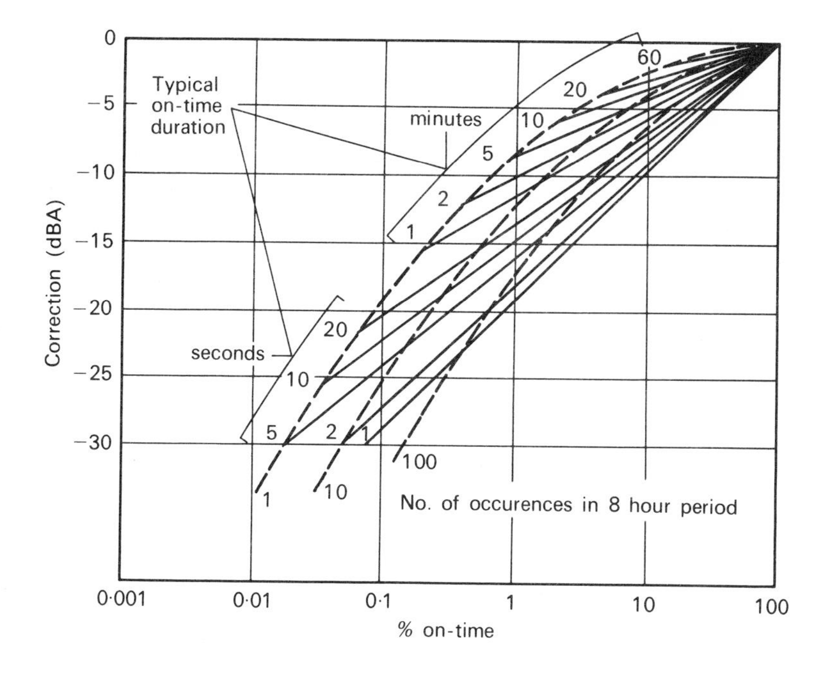


*Fig. 6.6* Intermittency and duration correction for other than night time

assessment. A possible investigator's report form is shown in Table 6·5. A suitable table for calculation of the criteria and corrected noise level is given in Table 6·6. The reader is referred to BS 4142 : 1967 for further details.

**Table 6·5**

NOISE INVESTIGATION REPORT

*Date* ______________________ *Name of Investigator* ______________________

*Site* ______________________________________________

*Type and No. of instrument* ______________________________

*Age of Investigator* __________ *Hearing defects* ______________________

*Investigator's assessment of noise (on first coming to site)*

1. General noise level ______________________________
2. Source of noise (traffic, trains, aircraft, machinery etc.)

______________________________________________

3. Is there a pure tone? ______________________________
4. If so, is it a high or low pitch? ______________________________
5. Is there any intermittent noise? ______________________________
6. Is there one particular noise source and what is its relationship to the background noise? ______________________________

*Assessment of noise by complainant and other occupiers*

General impression ______________________________

Any particularly annoying features ______________________________

*Complainant's statement*

______________________________________________

**Example 6·3**

A new factory 300 m long by 20 m wide is to be built on part of a disused railway line, in an area containing light industry. Existing houses are situated some 60 m from the side of the proposed new building. It is intended to operate the factory 24 hours per day throughout the year. The same manufacturer operates a smaller factory (in another area) of the same construction as that proposed, but with a quarter the number of identical machines. Measurements show that the sound pressure level at 7·5 m from the older factory is 60 dBA and there is a definite hum.

Find whether complaints would be expected if the proposed factory were constructed and if so, how much improvement in sound insulation would be needed. Background noise is 50 dBA.

| | |
|---|---|
| Estimated noise level at 7·5 m | = 60+6 |
| (four times the number of machines) | = 66 dBA |
| Expected noise level at 60 m | = 66−9 |
| (factory assumed to be a line source) | = 57 dBA |

The corrected noise level and corrected criterion are then calculated using Table 6·6.

**Table 6·6**

*Rating of Noise BS* 4142: 1967

| | Day time | | Night time | |
|---|---|---|---|---|
| | Old noise | New noise | Old noise | New noise |
| Measured noise level | | | | 57 dBA |
| Tonal character correction | | | | 5 |
| Impulsive character correction | | | | — |
| Intermittency and duration correction | | | | — |
| Corrected noise level | | | | 62 dBA (estimated) |
| Background noise level | | | | 50 dBA |
| Basic criterion | 50 | 50 | 50 | 50 |
| Correction for type of installation | | | | 0 |
| Correction for type of district | | | | +10 |
| Correction for time of day | | | | −5 |
| Correction for season | | | | 0 |
| Corrected criterion | | | | 55 |
| Corrected noise level minus background level | | | | +12 dBA |
| Corrected noise level minus corrected criterion | | | | 7 dBA |
| Conclusion | | | | Complaints possible |

It can be seen that the background noise is less than the corrected estimated level and is not actually needed. The corrected noise level is 7 dBA above the criteria and a few complaints could be expected. An improvement of 5–10 dB in the insulation should eliminate these.

## Motor Vehicle Noise Criteria

In 1963 the Committee on the Problem of Noise suggested that the maximum sound pressure levels for new vehicles should be:

| | |
|---|---|
| All vehicles except motor cycles | 85 dBA |
| Motor cycles | 90 dBA |

Legislation required that from 1 July 1968 all vehicles except those registered prior to 1931 must not exceed:

80 dBA for motor cycles under 50 cc
90 dBA for other motor cycles
92 dBA for heavy vehicles
87 dBA for passenger cars
88 dBA for light goods vehicles

In addition, from 1 April 1970 nearly all new vehicles have to be about 3 dB below these limits.

These restrictions have been introduced to try to prevent the continuous increase in the total traffic noise which has been about 1 dB per year.

Recent work by the Building Research Station suggests 'the traffic noise index' which could be used for town planning purposes. The traffic noise index (TNI) takes into account the levels of noise which are exceeded for 10 per cent and 90 per cent of the time during 24 hour periods. The 90 per cent level is the average, while the 10 per cent level gives an average maxima. The TNI is computed from:

$$\text{TNI} = 4 \times 10 \text{ per cent level} - 3 \times 90 \text{ per cent level } - 30$$

where both 10 per cent and 90 per cent levels are measured at 1 m from a building. Thirty is subtracted merely to yield a convenient numerical scale.

If measurements of the 10 per cent and 90 per cent levels are made well away from a building façade the measured values would be reduced by about 3 dB and the formula becomes:

$$\text{TNI} = 4 \times 10 \text{ per cent level } - 3 \times 90 \text{ per cent level } - 27$$

Accepting a TNI of 74 for a planning criteria as suggested, would mean that only one person in forty is likely to be dissatisfied. Using this index as the criteria the minimum distance of buildings from a motorway can be determined. As explained in Chapter 5 sound pressure levels decrease by 6 dB for an isolated source and by 3 dB for a line source each time the distance is doubled. It can be seen that TNI should decrease by 15 for each doubling of the distance. In practice it is slightly lower at about 14. If the TNI is measured at a certain distance $d_0$ from the road, the distance $d_1$ at which an acceptable TNI of 74 would be achieved can be calculated from the formula:

$$\text{Required reduction in TNI} = 45 \log_{10} \frac{d_1}{d_0}$$

### Example 6·4

The TNI at 7·5 metres from a roadway was found to be 104. What is the minimum distance from the road at which a building should be erected to achieve a TNI of 74 at its face?

Reduction in TNI = 104 − 74

$$\therefore \quad 30 = 45 \log_{10} \frac{d_1}{d_0}$$

$$\log_{10} \frac{d_1}{d_0} = 0{\cdot}6667$$

$$\therefore \quad \frac{d_1}{7{\cdot}5} = 4{\cdot}641$$

$$d_1 = 34{\cdot}8 \text{ metres}$$

Distance might be traded for sound insulation in the form of double glazing.

### Questions

(1) The recommended internal 10 per cent noise level for a courtroom is 35 dBA. What average insulation is required to achieve this if the octave band spectrum of traffic noise outside is:

| Frequency Hz | 63 | 125 | 250 | 500 | 1000 | 2000 | 4000 | 8000 |
|---|---|---|---|---|---|---|---|---|
| S.P.L. dB | 81 | 79 | 76 | 70 | 66 | 65 | 58 | 46 |

(2) A church is to be built near a busy road intersection at which the noise levels at weekends are:

| Frequency Hz | 63 | 125 | 250 | 500 | 1000 | 2000 | 4000 | 8000 |
|---|---|---|---|---|---|---|---|---|
| S.P.L. dB | 81 | 84 | 84 | 82 | 75 | 70 | 66 | 58 |

Find the insulation required to achieve NC 25 inside.

(3) An office is to be built in part of a factory and the insulation required is such that the SIL for a normal male voice at 1 m is achieved. The levels in the factory are:

| Centre frequency Hz | 63 | 125 | 250 | 500 | 1000 | 2000 | 4000 | 8000 |
|---|---|---|---|---|---|---|---|---|
| S.P.L. dB | 50 | 53 | 67 | 74 | 72 | 87 | 78 | 64 |

What average insulation is required at the SIL frequencies? If the insulation increases by 5 dB each time the frequency is doubled, what insulation is obtained at each frequency if it was just sufficient in the 1200–2400 Hz. band? Explain why the appropriate NC curve normally has an advantage over the SIL.

(4) A recording studio is situated in a street which is partly residential. It is intended to be able to continue recording sessions until late at night without disturbance to neighbours who may be sleeping. The bedrooms of local residents are fitted with double glazed windows giving the following insulation:

| Frequency Hz | 125 | 250 | 500 | 1000 | 2000 | 4000 | 8000 |
|---|---|---|---|---|---|---|---|
| Insulation dB | 19 | 19 | 27 | 37 | 47 | 52 | 53 |

If the highest level of sound at any frequency during a recording session is estimated to be 100 dB, what insulation is needed in the studio to achieve NC 30 level in surrounding houses?

(5) A small new factory is to be built in an urban area. The S.P.L. inside with all machines running is expected to be 105 dBA and is a continuous noise with no particular tonal characteristics. Domestic dwellings are situated close to one side of the factory. What is the minimum weight of wall to make it unlikely that there will be complaints from householders if:
(a) the factory is only in use during the day?
(b) the factory is to be used at night?
(Assume reduction in dB $= 10+14{\cdot}5 \log_{10} m$, where $m$ is mass/unit area in $kg/m^2$.)
If the factory was designed for use (a) and it was later changed to use (b), what area of material with an absorption coefficient of 0·8 would be needed? The volume of the factory is 1000 $m^3$ and its original reverberation time was 3·2 s.

(6) At a particular site 66 dBA was exceeded by road traffic for 90 per cent of the time and 81 dBA for 10 per cent of the time. The measurements were made 10 m from the roadway and clear of any obstructions. How far from the road should a building be erected so that the TNI at its face was 74?

# Answers

## Chapter 1

(1) 111 dB
(2) 101·5 dB
(3) 4
(4) 73·4 dB   $1{\cdot}995 \times 10^{-5}$ W/m$^2$
(5) 95·5 dB   74·3 dBA
(6) $4{\cdot}9 \times 10^{-8}$ W/m$^2$
(7) 20·25 N/m$^2$
(8) 70 to 77 dB re $10^{-12}$ W
(9) $2{\cdot}5 \times 10^{-4}$ W/m$^2$   84 dB
(10) 96 dB
(11) $6{\cdot}3 \times 10^{-2}$ W
(12) 0·18 W

## Chapter 2

(1) 88 phons
(2) 35 phons
(3) 90 PNdB
(4) Below 40 PNdB
(5) 106 phons
(6) 109 PNdB

## Chapter 3

(1) 4400
(2) 600 m$^2$   0·8 s   70 m$^2$   (approx.)
(3) 51
(4)

| 0·27 | 0·32 | 0·36 | 0·39 | 0·37 |
|---|---|---|---|---|
| (125) | (250) | (500) | (1000) | (2000) |
| Hz | Hz | Hz | Hz | Hz |

(5) Actual 1·87 s; optimum 1·63 s; 34 extra
(6) Optimum 1·2 s; actual 1·8 s
(7) 125 Hz — none; 250 Hz — 6·4 m$^2$; 500 Hz — 9 m$^2$; 1000 Hz — 10·3 m$^2$; 2000 Hz — 7·6 m$^2$; 4000 Hz — 4·3 m$^2$
(8) 0·87 s   54·3 m$^2$
(9) 3·27 s   4 dB
(10) Approx. 1800 m$^3$; R.T. about 1 s; approx. 50 m$^2$ abs
(11) Optimum R.T. 1·8 s; actual R.T. 2·0 s; 2·1 s
(12) 0·21

## Chapter 4

(1) 30 Hz
(2) 16 mm
(3) 2 m square
(4) 5810 N/mm$^2$
(5) Grade 2 (0·6 dB low); ISO impact protection margin = 0   ISO index = 65
(6) Grade 2
(7) 400 m/s

## Chapter 5

(1) 106·9 dB
(2) 58 m
(3) $8{\cdot}1 \times 10^6$ W
(4) 1·9 m$^2$
(5) 5 dB
(6) 36·5 dB
(7) 17, 17, 14, 13, 9, 8, 4 dB
(8) 1·6, 10, 14, 25, 29, 32, 29
(9) 27·5 dB
(10) 99 dB
(11) 13 dB
(12) 5 kg/m$^2$

## Chapter 6

(1) 39 dB
(2) 24, 37, 45, 50, 47, 45, 44, 37 dB
(3) 25 dB; 8, 13, 18, 23, 28, 33, 38, 43 dB
(4) 30, 38, 36, 31, 23, 20, 20
(5) 260 kg/m$^2$;  1268 kg/m$^2$  (impossibly  high); 562 m$^2$
(6) 36 m

# Bibliography


### Chapter 1

1. *Glossary of Acoustical Terms.* B.S. 661, 1955 (All British Standards and I.S.O. Recommendations may be obtained from the British Standards Institution, British Standards House, 2 Park Street, London, W1Y 4AA).
2. *Preferred Frequencies for Acoustical Measurements.* B.S. 3593:1963.
3. *Preferred Frequencies and Band Numbers for Acoustical Measurements.* U.S.A. Standard. S.I. 6, 1967.
4. *Specification for Octave and One Third Octave Band-Pass Filters,* B.S. 2475:1964.
5. *Octave, Half Octave and Third Octave Band Filter Sets.* U.S.A. Standard. S.I. 11, 1966.
6. 'Noise Measurement Techniques', *Notes on Applied Science No. 10. National Physical Laboratory,* Department of Scientific and Industrial Research.

### Chapter 2

1. *The Relations of Hearing Loss to Noise Exposure.* U.S.A. Standard. Z 24-X-2.
2. *Method for the Measurement of the Real-Ear Attenuation of Ear Protectors at Threshold.* U.S.A. Standard. Z 24.22:1957.
3. *Method for Calculating Loudness.* B.S. 4198:1967.
4. *The Relation Between the Sone Scale of Loudness and the Phon Scale of Loudness Level.* B.S. 3045:1958.
5. *Method of Calculating Loudness Level.* I.S.O., Recommendation 532, 1966.
6. *Expression of the Physical and Subjective Magnitudes of Sound or Noise.* I.S.O., Recommendation 131, 1959.
7. *Normal Equal-Loudness Contours for Pure Tones and Normal Threshold of Hearing under Free Field Listening Conditions.* I.S.O., Recommendation 226, 1961.
8. BURNS, WILLIAM. *Noise and Man.* John Murray.
9. *Noise and the Worker,* Safety, Health & Welfare, (New Series No. 25), H.M.S.O.
10. *Noise.* Final Report for the Committee on the Problem of Noise. H.M.S.O. Cmnd 2056:1966.
11. RODDA, MICHAEL. *Noise and Society.* Oliver & Boyd.
12. VAN BERGEIJK, W. A., PIERCE, J. R., and DAVID, E. E. *Waves and the Ear.* Heinemann.
13. MOLITOR, D. L. *Noise Abatement.* A Public Health Problem Report for Council of Europe, Strasbourg, 1964.

### Chapter 3

1. *Measurement of Absorption Coefficients in a Reverberation Room.* I.S.O., Recommendation 354, 1963.
2. DOELLE, L. L. *Acoustics in Architectural Design.* National Research Council Canada Division of Building Research Bibliography No. 29, 1965.
3. *Method for the Measurement of Sound Absorption Coefficients in a Reverberation Room.* B.S. 3638:1963.

4. EVANS, E. J. and BAZLEY, E. N. *Sound Absorbing Materials*. H.M.S.O.

5. PARKIN, P. H. and HUMPHREYS, H. R. *Acoustics, Noise and Buildings*. Faber.

6. PURKIS, H. J. *Building Physics: Acoustics*. Pergamon.

7. SCHROEDER, M. R. 'Natural Sounding Artificial Reverberation'. *Journal of the Audio-Engineering Society*, **10**, No. 3, July 1962, 219–23.

8. TAYLOR, H. O. 'Tube Method of Measuring Sound Absorption'. *Journal of the Acoustical Society of America*, **24**, no. 6, Nov. 1952.

9. *Handbook of Noise Control*. Edited by C. M. Harris. McGraw-Hill.

10. KNUDSEN, V. O. and HARRIS, C. M. *Acoustical Designing in Agriculture*. John Wiley & Sons.

11. MOORE, J. E. *Design for Good Acoustics*. Architectural Press.

12. BERANEK, L. L. *Music, Acoustics and Architecture*. McGraw-Hill, 1960.

13. *Proc. Phys. Soc.* 59, 535. Bate and Pillow, 1947.

**Chapters 4 and 5**

1. *Laboratory Measurement of Air-Borne Sound Transmission Loss of Building Floors and Walls*. U.S.A. Standard Z 24.19, 1957.

2. *Field and Laboratory Measurements of Airborne and Impact Sound Transmission*. I.S.O. Recommendation 140, 1960.

3. *Rating of Sound Insulation for Dwellings*. I.S.O., Recommendation 717, 1968.

4. BERANEK, L. L., *Noise Reduction*. McGraw-Hill.

5. *Handbook of Noise Control*. Edited by C. M. Harris. McGraw-Hill.

6. PARKIN, P. H. and HUMPHREYS, H. R. *Acoustics, Noise and Buildings*. Faber.

7. PURKIS, H. J. *Building Physics: Acoustics*. Pergamon.

8. *Sound Insulation of Traditional Dwellings 1 and 2 B.R.S.* Digests 102 and 103. March 1969. H.M.S.O.

9. 'Sound Insulation and New Forms of Construction'. *B.R.S. Digest*, 96, H.M.S.O.

10. BAZLEY, E. N., *The Airborne Sound Insulation of Partitions*. H.M.S.O.

11. *Measurement of Airborne and Impact Sound Transmission in Buildings*. B.S. 2750:1956.

12. HINES, W. A. *Noise Control in Industry*. Business Publications Ltd.

13. 'The Control of Noise'. *National Physical Laboratory Symposium, No. 12*. H.M.S.O.

14. KING, A. J. *The Measurement and Suppression of Noise*. Chapman & Hall.

15. MOORE, J. E. *Design for Noise Reduction*. Architectural Press.

16. CREMER, L. 'The Propagation of Structure-Borne Sound'. *DSIR Report 1*. Series B.

17. CREMER, L. 'Calculation of Sound Propagation in Structures'. *Acustica* **3**, 317, 1953.

18. JUNGER, M. C. 'Structure-borne Noise'. *1st. Symposium on Naval Structural Engineering*.

19. BERANEK, L. L. 'The Transmission and Radiation of Acoustic Waves by Structures'. The Forty-fifth Thomas Hawksley Lecture, *Proc. Inst. Mech. Engineers*, **173**, 1959.

20. *London Noise Survey*. Building Research Station, Ministry of Building and Works. H.M.S.O.

21. 'Noise in Factories'. *Factory Building Studies 6*. H.M.S.O., 1962.

22. 'Sound Insulation and Noise Reduction'. *British Standard Code of Practice* (*CP*: *3*), Chapter 3, 1960.

**Chapter 6**

1. *Background Noise in Audiometer Rooms*. U.S.A. Standard S3.1, 1960.

2. *The Normal Threshold of Hearing for Pure Tones by Earphone Listening*. B.S. 2497:1954.

3. *Pure Tone Audiometers*. B.S. 2980:1958.

4. *Standard Reference Zero for the Calibration of Pure Tone Audiometers*. I.S.O., Recommendation 389.

5. *Method for the Measurement of the Real-Ear Attenuation of Ear Protectors at Threshold*. U.S.A. Standard Z24.22, 1957.

6. KEIGHLEY, E. C. 'The Determination of Acceptability Criteria for Office Noise'. *Building Research Current Paper, Research Series 58*.

7. *Method of Rating Industrial Noise Affecting Mixed Residential and Industrial Areas*. B.S. 4142:1967.

8. ROWLANDS, E. *Noise in Hospitals: Subjective and Objective Criteria*. Environmental Design Research Unit, University College, London.

9. *Noise in Hospitals*. U.S. Department of Health Education and Welfare.

10. *Hospital Design*. Note 4, 'Noise Control'. Ministry of Health: London. H.M.S.O., 1966.

11. *Noise Control in Hospitals* (1) 1958 (2) 1960 King Edward's Hospital Fund for London, 34 King Street, E.C.2.

12. *Measurement of Noise Emitted by Motor Vehicles*. I.S.O. Recommendation 362, 1964.

13. *Method for the Measurement of Noise Emitted by Motor Vehicles*. B.S. 3425:1966.

14. GRIFFITHS, I. D., and LANGDON, F. J. 'Subjective Response to Road Traffic Noise'. *B.R.S. Current Paper 37/68*.

15. LANGDON, F. J., and SCHOLES, W. E., 'The Traffic Noise Index: A Method of Controlling Noise Nuisance'. *B.R.S. Current Paper 38/68*.

# Index